D0876307

TOPICS IN COMPLEX FUNCTION THEORY

INTERSCIENCE TRACTS IN PURE AND APPLIED MATHEMATICS

1. *D. Montgomery and L. Zippin*—**Topological Transformation Groups**
2. *Fritz John*—**Plane Waves and Spherical Means Applied to Partial Differential Equations**
3. *E. Artin*—**Geometric Algebra**
4. *R. D. Richtmyer*—**Difference Methods for Initial-Value Problems**
5. *Serge Lang*—**Introduction to Algebraic Geometry**
6. *Herbert Busemann*—**Convex Surfaces**
7. *Serge Lang*—**Abelian Varieties**
8. *S. M. Ulam*—**A Collection of Mathematical Problems**
9. *I. M. Gel'fand*—**Lectures on Linear Algebra**
10. *Nathan Jacobson*—**Lie Algebras**
11. *Serge Lang*—**Diophantine Geometry**
12. *Walter Rudin*—**Fourier Analysis on Groups**
13. *Masayoshi Nagata*—**Local Rings**
14. *Ivan Niven*—**Diophantine Approximations**
15. *S. Kobayashi and K. Nomizu*—**Foundations of Differential Geometry. In two volumes**
16. *J. Plemelj*—**Problems in the Sense of Riemann and Klein**
17. *Richard Cohn*—**Difference Algebra**
18. *Henry B. Mann*—**Addition Theorems: The Addition Theorems of Group Theory and Number Theory**
19. *Robert Ash*—**Information Theory**
20. *W. Magnus and S. Winkler*—**Hill's Equation**
21. *Zellig Harris*—**Mathematical Operations in Linguistics**
22. *C. Corduneanu*—**Almost-Periodic Functions**
23. *Richard S. Bucy and Peter D. Joseph*—**Filtering and Its Applications**
24. *Paulo Ribenboim*—**Rings and Modules** (in preparation)
25. *C. L. Siegel*—**Topics in Complex Function Theory**
    Vol. I Elliptic Functions and Uniformization Theory
    Vol. II Automorphic Functions and Abelian Integrals

TOPICS IN COMPLEX FUNCTION THEORY

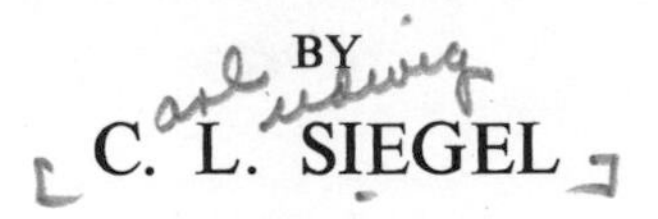

BY

C. L. SIEGEL

VOL. II

Automorphic Functions and Abelian Integrals

TRANSLATED FROM THE ORIGINAL GERMAN BY

A. SHENITZER

York University, Toronto

AND

M. TRETKOFF

Courant Institute of Mathematical Sciences, New York

WILEY–INTERSCIENCE

A DIVISION OF JOHN WILEY & SONS, INC.

NEW YORK • LONDON • SYDNEY • TORONTO

Library of Congress Catalog Card Number: 69-19931

ISBN 0-471-79080 X

Printed in the United States of America

10 9 8 7 6 5 4 3 2 1

Preface

Like Volume I, Volume II is based on the careful notes of my lectures prepared by Dr. E. Gottschling, then a young student. Compared to the original version, both chapters have been expanded in a number of places. Considering the paucity of relevant material, the new section on canonical polygons will, it is hoped, be helpful.

I wish to express my thanks to G. Köhler for his care in the preparation of the figures and in proofreading.

C. L. Siegel

Preface to the English Edition

Topics in Complex Function Theory consists of three parts corresponding to three consecutive courses taught by Professor Carl L. Siegel at the University of Göttingen. Lecture notes, in German, were taken by Erhard Gottschling and (for the second half of Part III) by Helmut Klingen. They were made available in mimeographed form by the Mathematical Institute of the University of Göttingen and were later revised by Professor Siegel, who then gave me permission to arrange for the publication of an English translation. The text was solicited for "Interscience Tracts" by Professor Lipman Bers. That Parts I and II can now appear is due to the dedicated efforts of Professor Abe Shenitzer who produced a complete first draft in English. He was assisted in questions of terminology and style and in checking the correctness of the translation by his colleagues Donald Solitar and Trueman MacHenry at York University and by Mr. Marvin Tretkoff at New York University. In some cases the translators have added notes concerning the use of certain terms, but these notes are marked as such.

A translation of Part III on "Abelian Functions and Modular Functions in Several Variables" is in preparation. After its completion an English text written by a great mathematician of our time will present one of the most fascinating fields of mathematics.

Wilhelm Magnus
New York University

New York
May 1970

Contents

TOPICS IN COMPLEX FUNCTION THEORY

3

Automorphic functions

1. Fractional linear transformations

Fractional linear transformations appeared in the last section of Chapter 2 (Volume I). On that occasion we studied some of their properties. We now continue this study with a view to establishing a connection between fractional linear transformations and noneuclidean geometry.

Let

$$w = \frac{az + b}{cz + d} \tag{1}$$

be a fractional linear transformation with complex coefficients a, b, c, d which are not all zero. For this transformation not to be constant it is necessary and sufficient that the determinant $ad - bc$ be different from zero. This condition will be assumed in the sequel. The inverse substitution

$$z = \frac{dw - b}{-cw + a} \tag{2}$$

has the same determinant. Composition of two substitutions

$$w = \frac{az + b}{cz + d}, \qquad z = \frac{\alpha s + \beta}{\gamma s + \delta} \qquad (ad - bc \neq 0; \quad \alpha\delta - \beta\gamma \neq 0)$$

yields a substitution

$$w = \frac{(a\alpha + b\gamma)s + (\alpha\beta + b\delta)}{(c\alpha + d\gamma)s + (c\beta + d\delta)}$$

with determinant

$$(a\alpha + b\gamma)(c\beta + d\delta) - (a\beta + b\delta)(c\alpha + d\gamma) = (ad - bc)(\alpha\delta - \beta\gamma) \neq 0.$$

It follows that the fractional linear substitutions with nonvanishing determinant form a group.

Theorem 1: The transformation (1) represents a single-valued one-to-one conformal mapping of the complex number sphere onto itself.

Proof: Expressions (1) and (2) show that the mapping is one-to-one and single-valued, because rational functions of one variable are meromorphic everywhere on the number sphere. For $c = 0$ and $z \neq \infty$, we have $dw/dz = a/d \neq 0$, so that the mapping from z to w is conformal for all finite z. For $c \neq 0$ and $z \neq -d/c$, we have $dw/dz = (ad - bc)/(cz + d)^2 \neq 0$, so that the mapping is again conformal. For $z = \infty$ we must introduce, in accordance with the definition of analyticity of a function in a neighborhood of infinity, the local parameter $t = z^{-1}$. Then, for $c \neq 0$, we obtain at $t = 0$ the derivative

$$\frac{dw}{dt} = \frac{bc - ad}{(c + dt)^2} = \frac{bc - ad}{c^2} \neq 0.$$

When $c = 0$, the point $z = \infty$ is mapped to $w = \infty$; and when $c \neq 0$, the point $-d/c$ is mapped to $w = \infty$. Thus we have to replace w by the local parameter

$$u = w^{-1} = \frac{cz + d}{az + b} = \frac{dt + c}{bt + a}.$$

Then we have

$$\frac{du}{dt} = \frac{ad - bc}{(bt + a)^2} = \frac{d}{a} \neq 0, \qquad \text{when} \quad c = 0, \quad t = 0,$$

and

$$\frac{du}{dz} = \frac{bc - ad}{(az + b)^2} = \frac{c^2}{bc - ad} \neq 0, \qquad \text{when} \quad c \neq 0, \quad z = -\frac{d}{c}.$$

It follows that the mapping is conformal without exception.

The mapping (1) is uniquely determined by assigning the four constants a, b, c, d. On the other hand, the mapping determines the four coefficients only up to a nonzero common factor. Specifically, if

$$\frac{az + b}{cz + d} = \frac{\alpha z + \beta}{\gamma z + \delta} \, (ad - bc \neq 0; \qquad \alpha\delta - \beta\gamma \neq 0)$$

identically in z, then, comparing coefficients, we obtain the relations

$$a\gamma = c\alpha, \qquad a\delta + b\gamma = \alpha d + \beta c, \qquad d\beta = b\delta.$$

The first and third of these conditions yield

$$\alpha = pa, \qquad \gamma = pc, \qquad \beta = qb, \qquad \delta = qd.$$

Upon substitution in the middle relation, we obtain

$$q(ad - bc) = p(ad - bc).$$

Thus $p = q$, and our assertion follows.

If we denote the transformation (1) by $\mathfrak{T}$, then there is associated with $\mathfrak{T}$ not only the coefficient matrix

$$T = \begin{pmatrix} a & b \\ c & d \end{pmatrix},$$

but also all matrices qT obtained by multiplying T by a nonzero scalar factor q. This explains the fact that if we use (2) to represent the inverse transformation $\mathfrak{T}^{-1}$, then we obtain the coefficient matrix

$$\begin{pmatrix} d & -b \\ -c & a \end{pmatrix} = (ad - bc)T^{-1},$$

and not T^{-1}. If we normalized all fractional linear substitutions by imposing the condition $ad - bc = 1$, then we would obtain just the matrix T^{-1} in the expression (2) for $\mathfrak{T}^{-1}$. This normalization notwithstanding, there would still be associated with every transformation $\mathfrak{T}$ two matrices, namely T and $-T$. That is why we dispense, for the time being, with normalization of the determinant.

If we apply simultaneously a fractional linear substitution $\mathfrak{C}$ to both variables w and z of the substitution $\mathfrak{T}$ by putting

$$w = \frac{\alpha r + \beta}{\gamma r + \delta}, \qquad z = \frac{\alpha s + \beta}{\gamma s + \delta} (\alpha\delta - \beta\gamma \neq 0),$$

then r is obtained from s by a fractional linear substitution $\mathfrak{U}$ which is related to $\mathfrak{T}$ by $\mathfrak{U} = \mathfrak{C}^{-1}\mathfrak{T}\mathfrak{C}$. The substitutions $\mathfrak{U}$ and $\mathfrak{T}$ are said to be *conjugate*.

As is well known, the expression

$$D(z_1, z_2, z_3, z_4) = \frac{z_1 - z_3}{z_1 - z_4} \Big/ \frac{z_2 - z_3}{z_2 - z_4} = \frac{(z_1 - z_3)(z_2 - z_4)}{(z_1 - z_4)(z_2 - z_3)}$$

is called the *cross ratio* of the ordered sequence of four distinct points z_1, z_2, z_3, z_4. Here the value $z_k = \infty$ is admissible. In the latter case the expression for the cross ratio is made meaningful by substituting $z_k = t_k^{-1}$ in the fraction, simplifying the expression, and finally putting $t_k = 0$.

Theorem 2: The cross ratio is invariant under fractional linear transformations.

Proof: With z_k variable, we put

$$w_k = \frac{az_k + b}{cz_k + d} \qquad (k = 1, \ldots, 4; \quad ad - bc \neq 0)$$

and obtain

$$w_k - w_l = \frac{az_k + b}{cz_k + d} - \frac{az_l + b}{cz_l + d} = \frac{(ad - bc)(z_k - z_l)}{(cz_k + d)(cz_l + d)} \qquad (k, l = 1, \ldots, 4).$$

Using this relation for the four pairs of values—$k, l = 1, 3;\ 1, 4;\ 2, 3;\ 2, 4$—we obtain

$$D(w_1, w_2, w_3, w_4) = D(z_1, z_2, z_3, z_4).$$

This equality holds, to begin with, for arbitrary finite fixed values of the z_k, provided that they are distinct and the w_k are finite. To check the validity of this equality for the value infinity, we begin with finite values and, making use of continuity, pass to the limit.

By replacing z_1, z_2, z_3, z_4 with z, z_1, z_2, z_3 and supposing z variable, we obtain from Theorem 2 the formula

$$\frac{w - w_2}{w - w_3} \Big/ \frac{w_1 - w_2}{w_1 - w_3} = \frac{z - z_2}{z - z_3} \Big/ \frac{z_1 - z_2}{z_1 - z_3}. \tag{3}$$

This is a linear equation for the unknown w which, when solved for w, must lead back to the fractional linear substitution (1). Conversely, if we assign to three distinct points z_1, z_2, z_3 three arbitrary distinct image points w_1, w_2, w_3, then (3) defines a fractional linear transformation from z to w which carries z_k to w_k $(k = 1, 2, 3)$. Thus every fractional linear transformation is uniquely determined by prescribing three points and their images.

We now classify the fractional linear transformations in terms of their fixed points. The fixed points of (1) are defined by the condition

$$z = \frac{az + b}{cz + d},$$

which is equivalent to

$$cz^2 + (d - a)z - b = 0.$$

All the coefficients of this quadratic equation vanish if and only if (1) is the identity transformation; in this case, obviously, all z are fixed points. We now rule out this case. To admit ∞ as a possible fixed point, we introduce *homogeneous coordinates* by means of the substitution $z = \zeta_1\zeta_2^{-1}$, where ζ_1 and ζ_2 cannot vanish simultaneously. Then the fixed points of (1) are given by the nontrivial solutions ζ_1, ζ_2 of the two homogeneous linear equations

$$a\zeta_1 + b\zeta_2 = \lambda\zeta_1, \qquad c\zeta_1 + d\zeta_2 = \lambda\zeta_2,$$

or, equivalently, by the equations

$$(a - \lambda)\zeta_1 + b\zeta_2 = 0, \qquad c\zeta_1 + (d - \lambda)\zeta_2 = 0. \tag{4}$$

The values of λ for which the system (4) has nontrivial solutions are the roots of the quadratic equation

$$\begin{vmatrix} a - \lambda & b \\ c & d - \lambda \end{vmatrix} = \lambda^2 - (a + d)\lambda + ad - bc = 0. \tag{5}$$

These λ are called the *eigenvalues* of the matrix

$$T = \begin{pmatrix} a & b \\ c & d \end{pmatrix}.$$

We note that transition from T to qT requires replacement of λ by $q\lambda$. Since $ad - bc \neq 0$, the eigenvalues are different from zero. Now we distinguish two cases. In the first case T has two distinct eigenvalues λ_1 and λ_2. In the second case T has a single eigenvalue λ_0 which is counted twice. In the latter case the transformation $\mathfrak{T}$ is said to be *parabolic*.

If we replace λ in equation (4) by an eigenvalue of T, then, bearing in mind that not all four coefficients are zero, we can determine the ratio ζ_1/ζ_2 uniquely. If two distinct eigenvalues λ_1 and λ_2 determined the same solution ζ_1, ζ_2, we would obtain from (4) by subtraction the equations

$$(\lambda_2 - \lambda_1)\zeta_1 = 0, \qquad (\lambda_2 - \lambda_1)\zeta_2 = 0,$$

and these have only the trivial solution $\zeta_1 = \zeta_2 = 0$, which is a contradiction. It follows that in the case of distinct eigenvalues there are two distinct fixed points ρ_1 and ρ_2, whereas in the case of a single eigenvalue of multiplicity two, there is only one fixed point. In either case the point ∞ may appear as a fixed point.

To investigate further the first case, we fix an auxiliary point z_0 distinct from ρ_1 and ρ_2, and we let z be a variable point. The images of z_0 and z under our transformation are denoted by w_0 and w respectively. In view of Theorem 2, we have

$$\frac{w - \rho_1}{w - \rho_2} \Big/ \frac{w_0 - \rho_1}{w_0 - \rho_2} = \frac{z - \rho_1}{z - \rho_2} \Big/ \frac{z_0 - \rho_1}{z_0 - \rho_2}.$$

If we set $\mu = D(w_0, z_0, \rho_1, \rho_2)$, then

$$\frac{w - \rho_1}{w - \rho_2} = \mu \frac{z - \rho_1}{z - \rho_2}, \tag{6}$$

so that

$$D(w, z, \rho_1, \rho_2) = \mu. \tag{7}$$

Here we note that w_0 is distinct from z_0, ρ_1, ρ_2, and that (7) remains valid even if one of the four points w_0, z_0, ρ_1, ρ_2 is ∞. The number μ is different from $0, 1, \infty$, and it is defined independently of z and w. Thus (7) shows that the cross ratio $D(w, z, \rho_1, \rho_2)$ has the same value for all choices of $z \neq \rho_1, \rho_2$. Conversely, if we start with (7), where μ is a constant different from $0, 1, \infty$, and ρ_1 and ρ_2 are two distinct points, then w can be expressed via (6) as a fractional linear transformation with fixed points ρ_1 and ρ_2. It follows that (7) is a general expression for all such fractional linear transformations $\mathfrak{T}$. Notice that we arrive at the same substitution $\mathfrak{T}$ if we interchange ρ_1 and ρ_2 and replace μ by μ^{-1}.

The case of two distinct fixed points is further subdivided as follows. The substitution $\mathfrak{T}$ is called *hyperbolic* if μ is a positive real number, *elliptic* if μ has absolute value one, and *loxodromic* in all other cases.

Theorem 3: The substitution $\mathfrak{T}$ and its conjugate $\mathfrak{C}^{-1}\mathfrak{T}\mathfrak{C}$ are invariably of the same type (parabolic, hyperbolic, elliptic, or loxodromic).

Proof: Let C be a matrix associated with $\mathfrak{C}$ and let E be the 2×2 unit matrix. Since

$$C^{-1}(T - \lambda E)C = C^{-1}TC - \lambda E,$$

it follows that the matrices $T - \lambda E$ and $C^{-1}TC - \lambda E$, in the variable λ, have the same determinant. On the other hand, $C^{-1}TC$ is a matrix associated with the substitution $\mathfrak{C}^{-1}\mathfrak{T}\mathfrak{C}$, so that, by (5), $\mathfrak{C}^{-1}\mathfrak{T}\mathfrak{C}$ and $\mathfrak{T}$ have precisely the same eigenvalues. This means that $\mathfrak{C}^{-1}\mathfrak{T}\mathfrak{C}$ is parabolic if and only if $\mathfrak{T}$ is parabolic. Now assume that $\mathfrak{T}$ is not parabolic, and let w_0, z_0, ρ_1, ρ_2 be the images of w_0^*, z_0^*, ρ_1^*, ρ_2^* under the transformation $\mathfrak{C}$. Then ρ_1^* and ρ_2^* are fixed points of $\mathfrak{C}^{-1}\mathfrak{T}\mathfrak{C}$, and this transformation maps z_0^* to w_0^*. Theorem 2 implies that μ is also invariant under the transition from $\mathfrak{T}$ to $\mathfrak{C}^{-1}\mathfrak{T}\mathfrak{C}$. Our assertion follows.

We demonstrate the invariance of μ once again, but in a somewhat different fashion, by expressing μ in terms of the eigenvalues λ_1 and λ_2. If we introduce the matrix

$$B = \begin{pmatrix} 1 & -\rho_1 \\ 1 & -\rho_2 \end{pmatrix},$$

then using an appropriate scalar factor q, we obtain from expression (6) the relation

$$B^{-1}\begin{pmatrix} \mu & 0 \\ 0 & 1 \end{pmatrix}B = qT. \tag{8}$$

Thus qT has the eigenvalues μ and 1; and, using the appropriate subscripts, we have

$$\mu = \frac{\lambda_1}{\lambda_2}.$$

This also shows the reasonableness of the definition $\mu = 1$ in the parabolic case, since in that case equation (5) leads to the double eigenvalue λ_0. At all times we have

$$\mu + \mu^{-1} = \frac{\lambda_1^2 + \lambda_2^2}{\lambda_1\lambda_2} = \frac{(\lambda_1 + \lambda_2)^2}{\lambda_1\lambda_2} - 2 = \frac{(a + d)^2}{ad - bc} - 2.$$

From now on we disregard the loxodromic case and give a geometric discussion of the mapping in the remaining cases. In the hyperbolic and elliptic cases it is convenient to shift the fixed points ρ_1 and ρ_2 to 0 and ∞ by applying the similarity transformation

$$r = \frac{w - \rho_1}{w - \rho_2}, \qquad s = \frac{z - \rho_1}{z - \rho_2};$$

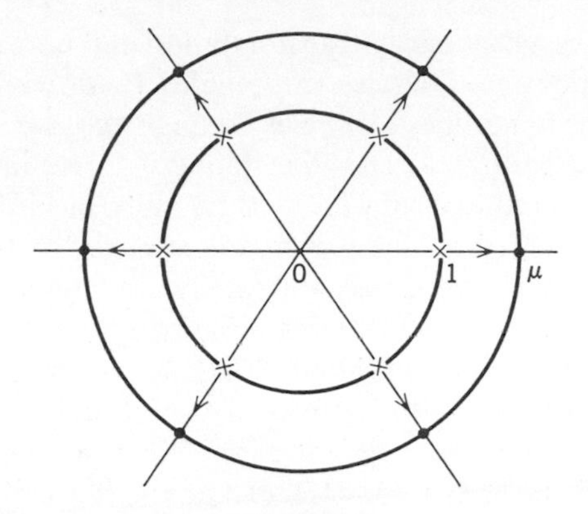

Figure 1

here we must replace both numerators or both denominators by 1 according as $\rho_1 = \infty$ or $\rho_2 = \infty$. If we write w and z for r and s, then, in accordance with (8), expression (6) takes the simple form

$$w = \mu z. \tag{9}$$

It follows that, in the hyperbolic case, every line through the origin is stretched by the positive factor μ. The orthogonal trajectories of this pencil of lines are the concentric circles about the origin. Each of these is transformed into a concentric circle whose radius is obtained by multiplying the given radius by μ (Figure 1). In the elliptic case we can put $\mu = e^{i\varphi}(0 < \varphi < 2\pi)$. Then the mapping (9) represents a rotation about the origin through the angle φ, so that every line through the origin is transformed into another line through the origin and every circle orthogonal to these lines is transformed into itself (Figure 2). If we now reintroduce the original variables

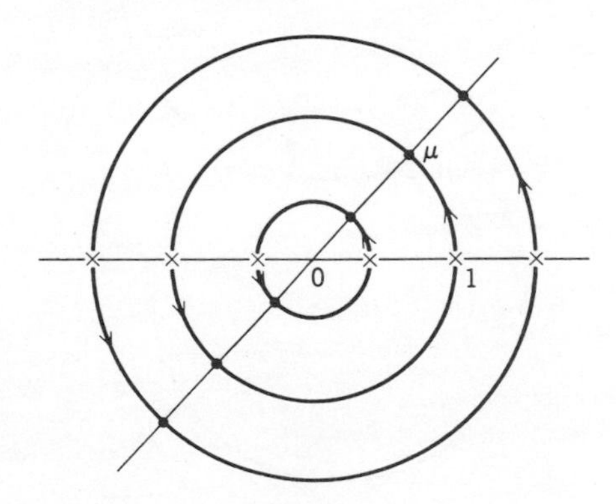

Figure 2

by means of the inverse similarity transformation and note that the fractional linear transformations take circles into circles (here we employ the usual convention of classifying lines as circles through ∞), we can conclude that in both cases under consideration the mapping $\mathfrak{T}$ leaves invariant the family of circles through ρ_1 and ρ_2 as well as the family of circles orthogonal to this family. In the hyperbolic case every circle of the first family is mapped onto itself, and the circles of the second family are interchanged in a definite manner which preserves their ordering (Figure 3). In the elliptic case the situation is reversed (Figure 4). Equation (7) shows how the circles invariant under $\mathfrak{T}$ are transformed into themselves (the value μ of the cross ratio of the four points w, z, ρ_1, ρ_2 is independent of z). We now want to establish that there are no circles invariant under $\mathfrak{T}$ except for those already considered. The equation of a line or a circle can be put in the form

$$\alpha z\bar{z} + \beta z + \bar{\beta}\bar{z} + \gamma = 0, \tag{10}$$

with α, γ real numbers. It follows that $\beta\bar{\beta} > \alpha\gamma$. Invariance under the mapping $w = \mu z$ implies that the following conditions must be satisfied for some appropriate factor q:

$$\alpha\mu\bar{\mu} = q\alpha, \qquad \beta\mu = q\beta, \qquad \bar{\beta}\bar{\mu} = q\bar{\beta}, \qquad \gamma = q\gamma.$$

If $\beta \neq 0$, it follows that $q = \mu = \bar{\mu} \neq 1$ and $\alpha = \gamma = 0$. Thus we are in the hyperbolic case and equation (10) represents a line in the pencil of lines

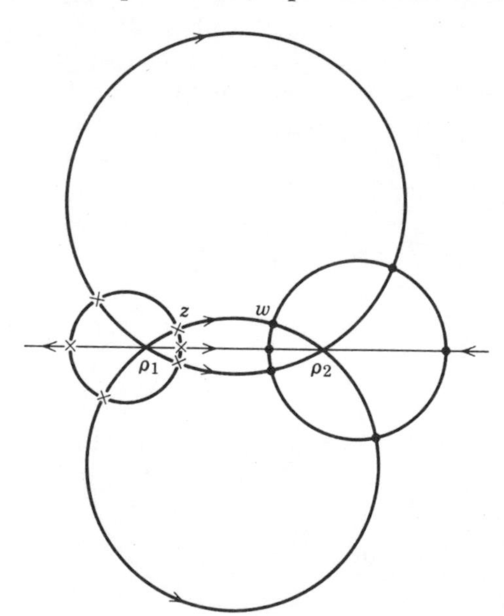

Figure 3

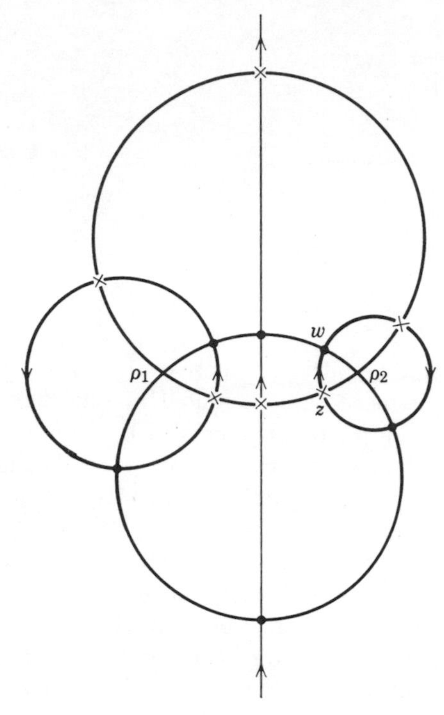

Figure 4

through 0. If $\beta = 0$, then $\gamma\alpha < 0$ and $\mu\bar{\mu} = q = 1$, so that we are in the elliptic case and (10) represents one of the concentric circles with center at the origin.

Now we consider the parabolic case and suppose that ∞ is the fixed point. Then we have $c = 0$ and $d \neq 0$ in (1). Furthermore, we have $a = d$; for, otherwise, there would be a finite fixed point in addition to the fixed point ∞. This means that

$$w = z + b \qquad (b \neq 0) \tag{11}$$

is the general expression for a parabolic substitution with the fixed point ∞. Equation (11) represents a translation by the vector b which carries every line parallel to b into itself and which permutes among themselves the members of any other family of parallel lines (Figure 5). If the fixed point ρ is different from ∞, then we move it to ∞ by applying to $\mathfrak{T}$ the similarity transformation

$$r = \frac{1}{w - \rho}, \qquad s = \frac{1}{z - \rho}.$$

Then the lines parallel to b correspond to a pencil of circles touching each

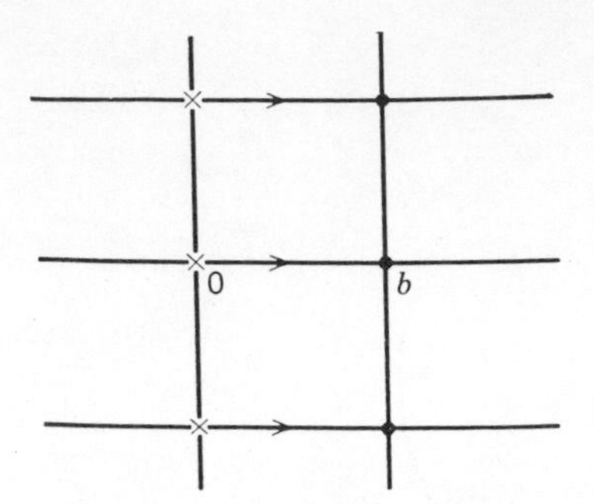

Figure 5

other at ρ and invariant under $\mathfrak{T}$, whereas the circles passing through ρ and orthogonal to this pencil are permuted among themselves (Figure 6). It is obvious that, except for the members of the family of lines parallel to b, no circle is invariant under the translation (11). This then is a geometric way of seeing that the parabolic case may be viewed as a common limiting case of the hyperbolic and elliptic cases.

Fractional linear substitutions $\mathfrak{T}$ whose square is the identity play a special role in the class of all fractional linear substitutions. These substitutions are called *involutions*. They can also be characterized by the fact that they coincide with their inverses $\mathfrak{T}^{-1}$. We assume that $\mathfrak{T}$ itself is not the identity. $\mathfrak{T}^2$ is the identity if and only if $T^2 = qE$ for some scalar $q \neq 0$. Since the squares of the eigenvalues λ of $\mathfrak{T}$ are the eigenvalues of $\mathfrak{T}^2$, we must have $\lambda^2 = q$. On the other hand, using (11), we see that a parabolic substitution always differs from its inverse. This means that an involution is never parabolic, and T must have two distinct eigenvalues λ_1 and λ_2 connected

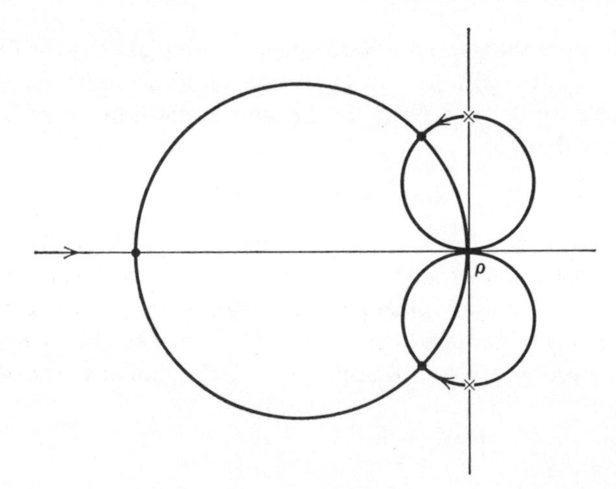

Figure 6

by the relations $\lambda_1^2 = \lambda_2^2$, $\lambda_1 = -\lambda_2$ and $\mu = \lambda_1/\lambda_2 = -1$. Conversely, if the condition $\mu = -1$ holds for the transformation $\mathfrak{T}$, then, by (7), the pairs of points w, z and ρ_1, ρ_2 form a harmonic set.* It follows that z is the image of w under $\mathfrak{T}$, and $\mathfrak{T}^2$ is therefore the identity. Thus we see that involutions are characterized by the condition $\mu = -1$. It is also clear that every similarity transformation carries an involution into another involution.

The following theorem was proved in part in the last section of Chapter 2 (Volume I).

Theorem 4: Let $\mathfrak{S}$ and $\mathfrak{T}$ be two fractional linear substitutions different from the identity. Then the relation $\mathfrak{S}\mathfrak{T} = \mathfrak{T}\mathfrak{S}$ holds if and only if $\mathfrak{S}$ and $\mathfrak{T}$ have the same fixed points or $\mathfrak{S}$ and $\mathfrak{T}$ are involutions whose pairs of fixed points form a harmonic set.

Proof: The statement of the theorem is not affected by simultaneous application of a similarity transformation to $\mathfrak{S}$ and $\mathfrak{T}$. Hence, when $\mathfrak{S}$ is not parabolic, we may assume that $\mathfrak{S}$ is in the normal form $w = \mu z$ and $\mathfrak{T}$ is in the general form (1). Then the assumed commutativity of $\mathfrak{S}$ and $\mathfrak{T}$ implies the following identity in z:

$$\mu \frac{az + b}{cz + d} = \frac{a\mu z + b}{c\mu z + d} \qquad (\mu \neq 0, 1; \quad ad - bc \neq 0).$$

Equating coefficients, we obtain

$$ac\mu^2 = ac\mu, \quad ad\mu + bc\mu^2 = ad\mu + bc, \quad bd\mu = bd,$$

so that

$$ac = 0, \qquad bc(\mu + 1) = 0, \qquad bd = 0.$$

If $d \neq 0$, then $b = 0$ and $c = 0$, so that $\mathfrak{T}$ also has the normal form $w = \nu z (\nu \neq 0, 1)$, where $\nu = ad^{-1}$. Transformations of this form commute with $\mathfrak{S}$ and have the same fixed points 0, ∞ for all ν. Conversely, if $\mathfrak{S}$ and $\mathfrak{T}$ have the same fixed points, then $\mathfrak{T}$ has this form. If $d = 0$, then $a = 0$, $\mu = -1$ and, putting $bc^{-1} = \rho^2$, $\mathfrak{T}$ has the form $w = \rho^2 z^{-1}$, while $\mathfrak{S}$ has the special form $w = -z$. Then, for every $\rho \neq 0$, $\mathfrak{S}$ and $\mathfrak{T}$ are commuting involutions and, since $D(0, \infty, \rho, -\rho) = -1$, their fixed points 0, ∞ and ρ, $-\rho$ form a harmonic set. Conversely, if $\mathfrak{S}$ and $\mathfrak{T}$ are two involutions whose pairs of fixed points 0, ∞ and ρ_1 ρ_2 form a harmonic set, then $\mu = -1$ and

$$-1 = D(0, \infty, \rho, \rho_2) = \frac{-\rho}{-\rho_2}.$$

* The pairs of points a, b and c, d are said to form a *harmonic set* if their cross ratio $D(a, b, c, d) = -1$. (Tr.)

Thus $\rho_2 = -\rho$. Since $\mathfrak{T}$ is an involution, (6) yields for it the formula

$$\frac{w-\rho}{w+\rho} = -\frac{z-\rho}{z+\rho},$$

or

$$w = \rho^2 z^{-1}.$$

Hence $\mathfrak{T}$ and $\mathfrak{S}$ commute.

It remains to consider the case of a parabolic $\mathfrak{S}$. Then, by (10), we may suppose $\mathfrak{S}$ in the normal form $w = z + \gamma$ $(\gamma \neq 0)$, while $\mathfrak{T}$ is again given by (1). Now we must have the identity

$$\frac{az+b}{cz+d} + \gamma = \frac{a(z+\gamma)+b}{c(z+\gamma)+d} \qquad (\gamma \neq 0, \quad ad - bc \neq 0).$$

Thus, equating coefficients, we have

$$(a + c\gamma)c = ac,$$
$$(a + c\gamma)(c\gamma + d) + (b + d\gamma)c = ad + (a\gamma + b)c,$$
$$(b + d\gamma)(c\gamma + d) = (a\gamma + b)d.$$

Therefore $c = 0$, $a = d$, so that $\mathfrak{T}$ must be parabolic with the same type of normal form as $\mathfrak{S}$ and with ∞ as the fixed point. Conversely, if $\mathfrak{T}$ has ∞ as its only fixed point, then it must be of the form $w = z + \delta$, and $\mathfrak{S}$ and $\mathfrak{T}$ commute. This completes the proof of our theorem.

We make use of some results of previous sections without repeating their proofs. A meromorphic function $w = f(z)$ effects a conformal mapping of the number sphere onto itself if and only if it is a fractional linear transformation, that is, if and only if

$$f(z) = \frac{az+b}{cz+d} \qquad (ad - bc \neq 0),$$

where we admit the possibility $c = 0$. An analytic function $w = f(z)$ effects a conformal mapping of the plane onto itself if and only if it is linear, that is, if and only if

$$f(z) = az + b \quad (a \neq 0).$$

An analytic function effects a conformal mapping of the open unit disk onto itself if and only if it is of the form

$$w = \lambda \frac{z-a}{1-\bar{a}z} \qquad (|\lambda| = 1, \quad |a| < 1). \tag{12}$$

The number a in (12) is the point mapped onto the origin. To shed light on the significance of the other parameter λ, we expand w in powers of $z - a$,

$$w = \lambda \frac{z-a}{(1-\bar{a}a) - \bar{a}(z-a)} = \frac{\lambda}{1-a\bar{a}}(z-a) + \cdots.$$

If, for small values of $|z - a|$, we ignore the higher-order terms, then we see that a sufficiently small neighborhood of $z = a$ is mapped onto a neighborhood of $w = 0$. The mapping is, approximately, a dilation with coefficient $(1 - a\bar{a})^{-1}$ followed by a rotation through the angle $\arg \lambda$. For differentials at $z = a$ we have the relation $dw = [\lambda/(1 - a\bar{a})]\, dz$. It is clear that the mapping (12) is determined by prescribing the point a and the angle of rotation $\arg \lambda$.

We find it convenient to replace the parameters λ and a in (12) with two other parameters p and q. To begin with we put $p = \sqrt{\bar{\lambda}}$, $q = -p\bar{a}$ for some definite choice of the value of the square root. Then (12) becomes

$$w = \frac{\bar{p}z + \bar{q}}{qz + p}, \tag{13}$$

with $|p| = 1$, $|q| < 1$, so that $p\bar{p} - q\bar{q} > 0$. By multiplying p and q with the same real positive factor, we can attain the normalization $p\bar{p} - q\bar{q} = 1$ without affecting the transformation (13), (of course, p need not retain the absolute value one). Then we have

$$1 - w\bar{w} = \frac{1 - z\bar{z}}{|p + qz|^2}. \tag{14}$$

Furthermore, we note that the mapping (12) is conformal everywhere on the number sphere, and that it carries the exterior of the unit circle into itself.

The substitution

$$z = \frac{s - i}{s + i}, \qquad s = i\,\frac{1 + z}{1 - z}$$

maps the interior of the unit disk $|z| < 1$ onto the upper half of the s-plane. If we put

$$a = \frac{p + \bar{p}}{2} + \frac{q + \bar{q}}{2}, \qquad b = \frac{\bar{p} - p}{2i} + \frac{q - \bar{q}}{2i},$$

$$c = \frac{p - \bar{p}}{2i} + \frac{q - \bar{q}}{2i}, \qquad d = \frac{p + \bar{p}}{2} - \frac{q + \bar{q}}{2},$$

then a, b, c, d are real, and we have

$$\begin{pmatrix} i & i \\ -1 & 1 \end{pmatrix}\begin{pmatrix} \bar{p} & \bar{q} \\ q & p \end{pmatrix}\begin{pmatrix} 1 & -i \\ 1 & i \end{pmatrix} = 2i\begin{pmatrix} a & b \\ c & d \end{pmatrix}, \qquad ad - bc = p\bar{p} - q\bar{q},$$

$$p = \frac{a + d}{2} + i\,\frac{c - b}{2}, \qquad q = \frac{a - d}{2} + i\,\frac{b + c}{2}.$$

It follows that every analytic function which maps the upper half of the s-plane conformally onto itself is given by

$$r = \frac{as + b}{cs + d}, \tag{15}$$

with a, b, c, d real and $ad - bc = 1$.

Theorem 5: The fractional linear transformation

$$w = \frac{\bar{p}z + \bar{q}}{qz + p} \qquad (p\bar{p} - q\bar{q} = 1) \tag{16}$$

is never loxodromic. If it is not the identity, then it is elliptic, parabolic, or hyperbolic according as $(p + \bar{p})^2 < 4, = 4, > 4$. In the first case the fixed points are connected by a reflection in the unit circle, and in the other two cases the fixed points lie on the unit circle.

Proof: The eigenvalues of the substitution are the roots of the equation $\lambda^2 - (p + \bar{p})\lambda + 1 = 0$ [cf. (5)] with discriminant

$$\Delta = (p + \bar{p})^2 - 4.$$

Since the coefficients are real, the equation has two distinct real roots, two distinct conjugate complex roots, or one real double root according as $\Delta > 0, \Delta < 0$, or $\Delta = 0$. It follows that $\Delta = 0$ corresponds to the parabolic case. For $\Delta < 0$ the ratio of the two eigenvalues is a number with absolute value one. For $\Delta > 0$, this ratio is real and positive, since the product of the eigenvalues is 1. It follows that $\Delta < 0$ corresponds to the elliptic case, and $\Delta > 0$ to the hyperbolic case. Since the unit circle is invariant under the mapping, we conclude that in the parabolic and hyperbolic cases all the fixed points must lie on $|z| = 1$. In the elliptic case the unit circle is orthogonal to the pencil of circles determined by the fixed points ρ_1 and ρ_2 and, therefore, must separate them. By (16)

$$\bar{w}^{-1} = \frac{\bar{p}\bar{z}^{-1} + \bar{q}}{q\bar{z}^{-1} + p},$$

so that, in particular, if ρ_1 is a fixed point then so is $\bar{\rho}_1^{-1}$. It follows that in the elliptic case $\rho_2 = \bar{\rho}_1^{-1}$. This completes the proof.

2. *Noneuclidean geometry*

The conformal mappings effected by analytic functions preserve orientation. By way of supplementing the investigations of the previous section we consider conformal mappings of the number sphere onto itself which reverse orientation. One such mapping is $w = \bar{z}$; we denote it by $\mathfrak{U}$. If $\mathfrak{V}$ is another such mapping, then the composition of $\mathfrak{V}$ and $\mathfrak{U}$ yields a conformal orientation-preserving mapping $\mathfrak{V}\mathfrak{U} = \mathfrak{T}$ of the number sphere onto itself. It follows that $\mathfrak{T}$ is analytic and, therefore, fractional-linear. Since $\mathfrak{U}^2$ is the identity, we have $\mathfrak{V} = \mathfrak{T}\mathfrak{U}$, so that

$$w = \frac{a\bar{z} + b}{c\bar{z} + d} \qquad (ad - bc \neq 0)$$

is the most general conformal and orientation-reversing mapping of the number sphere onto itself. This result carries over in a natural way to the case of the z-plane and the unit circle $|z| < 1$ in place of the number sphere, for the mapping $w = \bar{z}$ also carries these regions onto themselves. It follows that the formulas

$$w = a\bar{z} + b \qquad (a \neq 0), \qquad w = \frac{\bar{p}\bar{z} + \bar{q}}{q\bar{z} + p} \qquad (p\bar{p} - q\bar{q} = 1)$$

yield all the orientation-reversing conformal mappings of the plane onto itself and the unit circle onto itself. The steps which led to (15) in Section 1 now lead to the formula

$$w = \frac{a\bar{z} - b}{c\bar{z} - d} \qquad (ad - bc = 1),$$

a, b, c, d real, which gives the totality of orientation-reversing conformal mappings of the upper half plane onto itself.

Theorem 1: Four points z_1, z_2, z_3, z_4 lie on a circle if and only if their cross ratio is real. The two pairs of points z_1, z_2 and z_3, z_4 on this circle separate each other if and only if their cross ratio is negative.

Proof: Let $\mathfrak{T}$ be the fractional linear transformation which carries z_2, z_3, z_4 into $1, 0, \infty$, and let z_0 denote the image of z_1 under $\mathfrak{T}$. Since $\mathfrak{T}$ preserves cross ratios and carries circles into circles, we have

$$D(z_1, z_2, z_3, z_4) = D(z_0, 1, 0, \infty) = z_0, \tag{1}$$

and the circle determined by z_2, z_3, z_4 goes over into the circle determined by $1, 0, \infty$, that is, the real axis. It follows that the four given points lie on a circle if and only if z_0, hence the cross ratio of the four points, is real. Furthermore, this cross ratio is negative if and only if z_0 is negative, that is, if and only if the two pairs of points z_0, 1 and 0, ∞ on the extended real axis separate each other. To complete the proof we note that $\mathfrak{T}$ maps the circular arc from z_3 to z_4 through z_2 into the positive real half axis traversed in the positive direction.

The case when the pairs of points z_1, z_2 and z_3, z_4 on the circle do not separate each other will be of particular importance in the sequel. In that case we keep z_2, z_3, z_4 fixed and move z_1 from z_3 to z_4 by way of z_2. Then the point z_0 moves from 0 to ∞ by way of 1 and, in view of (1), the cross ratio $D(z_1, z_2, z_3, z_4)$ takes on all the positive values; in particular, the value 1 comes up for $z_1 = z_2$. If z_1, z_2, z_3, z_4 are arranged cyclically on the circle, then the cross ratio is greater than 1. In the special case when z_3 and z_4 are end points of a diameter of the circle, we have (Figure 7):

$$D(z_1, z_2, z_3, z_4) = \frac{\tan \varphi_1}{\tan \varphi_2},$$

where φ_k $(k = 1, 2)$ denotes the angle between the two vectors $z_k - z_4$ and $z_3 - z_4$. Hence, in this case, the cross ratio has simple geometric significance.

After these preliminaries we are in a position to introduce the *Poincaré model of plane noneuclidean geometry*. Let $\mathfrak{E}$ be the interior of the unit circle in the z-plane. Then every point of $\mathfrak{E}$ is called a *noneuclidean point* and $\mathfrak{E}$ itself is called the *noneuclidean plane*. If L is a circle orthogonal to the boundary of $\mathfrak{E}$, then the part G of L in $\mathfrak{E}$ is called a *noneuclidean line*. It is natural to admit a diameter of $\mathfrak{E}$ as a noneuclidean line. It is clear that every noneuclidean line G separates the noneuclidean plane into two noneuclidean half planes with common boundary G. An open noneuclidean half plane results from the omission of the boundary line.

In euclidean geometry we study *isometries*, that is, the mappings which leave the distance between two points invariant. Orientation-preserving isometries of the plane can be represented as products of translations and rotations. The analytical form of orientation-preserving isometries of the plane is $w = az + b$ $(|a| = 1)$, and the analytical form of orientation-reversing isometries of the plane is $w = a\bar{z} + b$. Any isometry of the plane carries euclidean lines into euclidean lines. The corresponding mappings of the noneuclidean plane are given by

$$(2) \qquad w = \frac{\bar{p}z + \bar{q}}{qz + p}, \qquad w = \frac{\bar{p}\bar{z} + \bar{q}}{q\bar{z} + p} \qquad (p\bar{p} - q\bar{q} = 1).$$

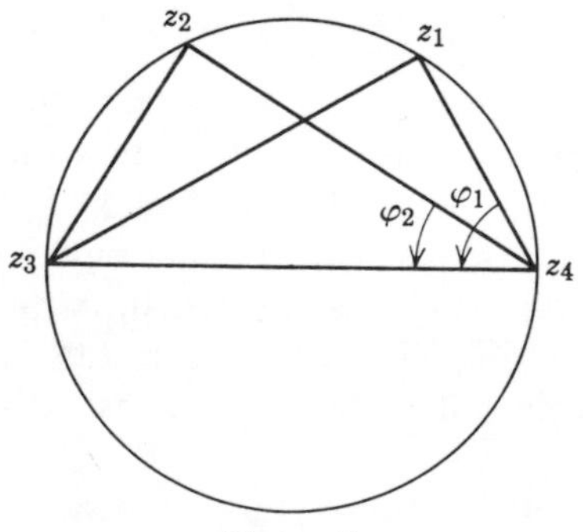

Figure 7

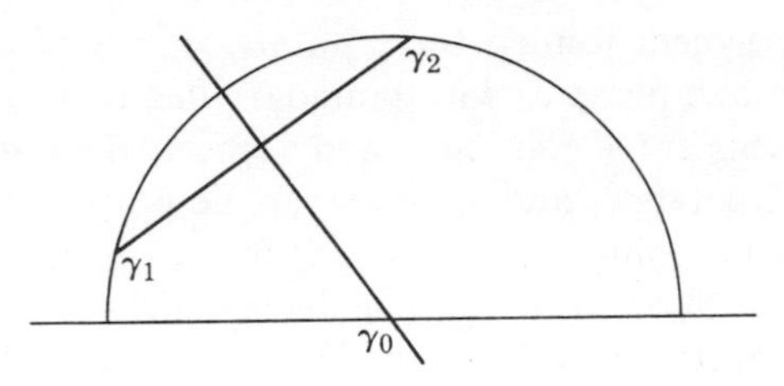

Figure 8

These mappings carry euclidean circles into euclidean circles and $\mathfrak{E}$ into itself and preserve or reverse angles. It follows that they carry noneuclidean lines into noneuclidean lines. We shall define the noneuclidean distance between two points so that it remains invariant under the transformations (2). The definition of our transformations shows that they form a group Ω^*; a fact which also admits of a simple direct proof. The mappings of the first kind, that is, the orientation-preserving mappings, are called *noneuclidean motions* and form an invariant subgroup Ω of index 2 in Ω^*.

Theorem 2: Two noneuclidean points determine a unique noneuclidean line.

Proof: Let z_1, z_2 be two distinct points in $\mathfrak{E}$. Let r_1, r_2 be the distinct images of z_1, z_2 under the mapping $r = i(1 + z)(1 - z)^{-1}$ of $\mathfrak{E}$ onto the upper half plane $\mathfrak{H}$. If r_1 and r_2 have unequal real parts, then the perpendicular bisector of the segment joining r_1 and r_2 meets the real axis in a point r_0 which is the center of a circle orthogonal to the real axis and is uniquely determined by r_1 and r_2 (Figure 8). If r_1 and r_2 have equal real parts, then the role of the circle orthogonal to the real axis is played by the parallel to the imaginary axis passing through points r_1 and r_2. If we map $\mathfrak{H}$ back onto $\mathfrak{E}$ by means of $z = (r - i)(r + i)^{-1}$ and bear in mind that this mapping preserves angles and carries circles (including lines) into circles, then we obtain a circle through z_1 and z_2 orthogonal to the boundary of $\mathfrak{E}$, and this circle is clearly unique. This completes the proof. Another proof is as follows. If there exists a circle through z_1 and z_2 orthogonal to the unit circle, then this circle must go into itself under the mapping $w = \bar{z}^{-1}$ and so must contain the points $\bar{z}_1^{-1}$ and $\bar{z}_2^{-1}$. Conversely, these four points determine the circle uniquely. Now, the cross ratio

$$D(z_1, z_2, \bar{z}_1^{-1}, \bar{z}_2^{-1}) = \frac{(z_1 - \bar{z}_1^{-1})(z_2 - \bar{z}_2^{-1})}{(z_1 - \bar{z}_2^{-1})(z_2 - \bar{z}_1^{-1})} = \frac{(1 - z_1\bar{z}_1)(1 - z_2\bar{z}_2)}{(1 - z_1\bar{z}_2)(1 - \bar{z}_1 z_2)}$$

is real, and our assertion follows from Theorem 1.

We call the part of the noneuclidean line through z_1 and z_2 lying between z_1 and z_2 the *noneuclidean segment with end points z_1 and z_2*. We call a subset of $\mathfrak{E}$ *noneuclidean-convex* if together with two of its points it contains the noneuclidean segment joining these points. If z_1 and z_2 lie in the same open noneuclidean half plane $\mathfrak{D}$ with boundary line G, then, by Theorem 2, the noneuclidean segment V joining z_1 and z_2 meets the noneuclidean line G at most once. The points z_1 and z_2, however, lie on the same side of G, so that V and G have no point in common. It follows that V lies in $\mathfrak{D}$, and we see that $\mathfrak{D}$ is noneuclidean-convex. The same conclusion holds if we adjoin G to $\mathfrak{D}$, that is, for every closed noneuclidean half plane.

We are now ready to define the noneuclidean distance from z_1 to z_2. To this end we consider the noneuclidean line determined by z_1 and z_2 and represented by the arc G in $\mathfrak{E}$ of a circle L orthogonal to the unit circle. We call the two end points z_3 and z_4 of G on $|z| = 1$ the *ends* of the noneuclidean line, and order them so that z_1, z_2, z_3, z_4 are arranged cyclically on L. Then we define

$$\delta(z_1, z_2) = \log D(z_1, z_2, z_3, z_4)$$

to be the *noneuclidean distance between* z_1 *and* z_2. By Theorem 1, the cross ratio of the four points is a real positive number and the logarithm is the real logarithm. In view of the invariance of the cross ratio under fractional linear transformations, the noneuclidean distance between two points is invariant under all noneuclidean motions. Since $D(z_1, z_2, z_3, z_4)$ is real, $\delta(z_1, z_2)$ remains invariant under the mapping $w = \bar{z}$ and therefore under the full group Ω^*.

It is clear that $\delta(z_1, z_2)$ depends on z_1 and z_2 alone, for z_3 and z_4 are uniquely determined by z_1 and z_2. Interchanging z_1 and z_2 requires interchanging z_3 and z_4, so that $\delta(z_1, z_2) = \delta(z_2, z_1)$. If z_1 and z_2 tend to a point z_0 in $\mathfrak{E}$, then the ends z_3 and z_4 need not tend to limiting positions on $|z| = 1$. Nevertheless, the cross ratio of the four points tends to 1 so that the noneuclidean distance from z_1 to z_2 tends to 0. It is therefore natural to define $\delta(z_1, z_2) = 0$ in the limiting case $z_1 = z_2$. We showed earlier that with the four points arranged on the circle in cyclic order, their cross ratio is greater than 1 and so $\delta(z_1, z_2) > 0$ for $z_1 \neq z_2$.

The euclidean distance $|z_1 - z_2|$ between two arbitrary points of the plane satisfies for every third point z_0 the triangle inequality

$$|z_1 - z_2| \leqslant |z_1 - z_0| + |z_0 - z_2|, \tag{3}$$

which states that the side of a triangle is less than the sum of the two remaining sides. In fact, for the equality sign to hold in (3), z_0 must lie on the segment between z_1 and z_2. We shall see that the noneuclidean distance $\delta(z_1, z_2)$ has the corresponding properties. At this point we prove only that

$$\delta(z_1, z_2) = \delta(z_1, z_0) + \delta(z_0, z_2)$$

for z_0 on the noneuclidean segment with end points z_1 and z_2 (Figure 9).

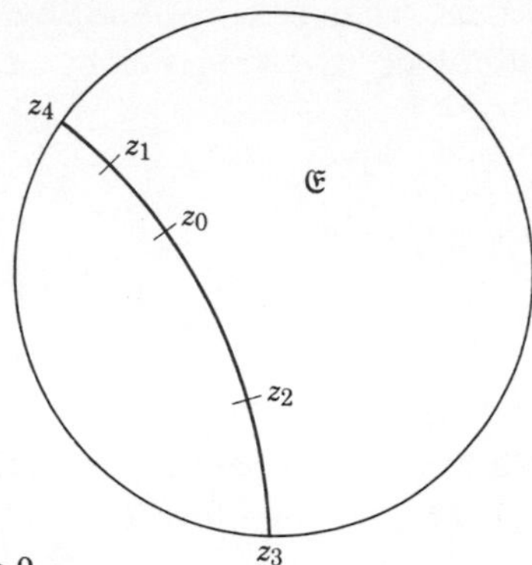

Figure 9

Note that

$$D(z_1, z_0, z_3, z_4) = \frac{(z_1 - z_3)(z_0 - z_4)}{(z_1 - z_4)(z_0 - z_3)},$$

$$D(z_0, z_2, z_3, z_4) = \frac{(z_0 - z_3)(z_2 - z_4)}{(z_0 - z_4)(z_2 - z_3)},$$

so that

$$D(z_1, z_0, z_3, z_4)D(z_0, z_2, z_3, z_4) = D(z_1, z_2, z_3, z_4).$$

Taking logarithms of both sides proves our assertion. This result justifies our measuring the noneuclidean distance between z_1 and z_2 by means of the logarithm of the cross ratio $D(z_1, z_2, z_3, z_4)$ rather than by means of the cross ratio itself.

It is possible to obtain for the distance between two points z_1 and z_2 an explicit expression which is a function of z_1 and z_2. To this end we choose a substitution in Ω which carries z_1 into the origin, z_2 into a real positive t, and, therefore, z_3 and z_4 into 1 and -1. Such a substitution is

$$w = \lambda \frac{z - a}{1 - \bar{a}z}, \qquad a = z_1, \qquad \lambda = \frac{|z_2 - z_1|\,(1 - \bar{z}_1 z_2)}{(z_2 - z_1)\,|1 - \bar{z}_1 z_2|}, \tag{4}$$

$$t = \frac{|z_2 - z_1|}{|1 - \bar{z}_1 z_2|}. \tag{5}$$

Then

$$\delta(z_1, z_2) = \delta(0, t) = \log D(0, t, 1, -1) = \log \frac{1 + t}{1 - t} = 2\left(t + \frac{t^3}{3} + \cdots\right). \tag{6}$$

This expression shows anew that if z_2 moves on the noneuclidean line with fixed ends z_3 and z_4 from z_1 to z_3, then $D(z_1, z_2, z_3, z_4)$ takes on all the positive values from 1 to ∞, and, correspondingly, the distance $\delta(z_1, z_2)$ takes on all the positive values from 0 to ∞. Furthermore,

$$|1 - \bar{z}_1 z_2|^2 - |z_2 - z_1|^2 = (1 - |z_1|^2)(1 - |z_2|^2),$$

$$|1 - \bar{z}_1 z_2| \geqslant 1 - |\bar{z}_1 z_2| \geqslant 1 - |z_1|$$

and (5) yields

$$1 - t^2 = \frac{(1 - |z_1|^2)(1 - |z_2|^2)}{|1 - \bar{z}_1 z_2|^2} \leqslant \frac{1 + |z_1|}{1 - |z_1|}(1 - |z_2|^2).$$

It follows that if $|z_2|$ approaches 1, with z_1 fixed and z_2 varying on $\mathfrak{E}$, then t approaches 1, and, in this transition to the limit, $\delta(z_1, z_2)$ also becomes infinite. On the other hand, (5) and (6) imply

$$\lim_{z_2 \to z_1} \frac{\delta(z_1, z_2)}{|z_2 - z_1|} = \lim_{z_2 \to z_1} \frac{2t}{|z_2 - z_1|} = \frac{2}{1 - |z_1|^2}. \tag{7}$$

In particular, for $z_1 = 0$, we see that the ratio of the noneuclidean to the euclidean distance of a point z_2 to the origin tends to the limit 2 as z_2 tends to 0.

The conjugation $\mathfrak{C}$ given by $z = (s - i)(s + i)^{-1}$ enables us to transplant the concepts of noneuclidean geometry developed for the unit disk $\mathfrak{E}$ to the upper half plane $\mathfrak{H}$. In that case the noneuclidean lines are the half circles and euclidean half lines in $\mathfrak{H}$ orthogonal to the real axis. If we define the noneuclidean distance $\delta(s_1, s_2)$ between two points s_1, s_2 in $\mathfrak{H}$ in analogy to the definition used in the noneuclidean geometry in $\mathfrak{E}$, then that distance is invariant under the corresponding noneuclidean motions given by the group $\mathfrak{C}^{-1}\Omega\mathfrak{C} = \Delta$, that is, under the transformations

$$r = \frac{as + b}{cs + d} \qquad (ad - bc = 1)$$

with real a, b, c, d. We also have invariance under the additional mappings

$$r = \frac{a\bar{s} - b}{c\bar{s} - a},$$

which, together with Δ, form the group $\mathfrak{C}^{-1}\Omega^*\mathfrak{C} = \Delta^*$. It is occasionally expedient to prove a theorem of noneuclidean geometry using the model $\mathfrak{H}$ and then carry it over to $\mathfrak{E}$.

Let us again start with two points z_1 and z_2 in $\mathfrak{E}$. If we use (4) to carry z_1 and z_2 by means of a noneuclidean motion to 0 and t, then there correspond to these points, under the mapping from $\mathfrak{E}$ to $\mathfrak{H}$, the points $s_1 = i$ and $s_2 = i(1 + t)(1 - t)^{-1}$ on the positive imaginary half axis. With $s_2 = i(1 + u)$, (5) and (6) yield the relation

$$\delta(z_1, z_2) = \delta(s_1, s_2) = \log(1 + u) = u - \frac{u^2}{2} + \cdots.$$

It follows that if a point s of $\mathfrak{H}$ converges to i, then the ratio of the noneuclidean to the euclidean distance from s to i tends to 1.

The invariance of noneuclidean distance under Ω^* can be verified by direct computation using the expression (5) for t as a function of z_1 and z_2. We shall use this approach to verify the triangle inequality, that is, to show that for three distinct points z_1, z_0, z_2 of $\mathfrak{E}$ we have invariably

$$\delta(z_1, z_2) \leqslant \delta(z_1, z_0) + \delta(z_0, z_2),$$

where the equality sign holds only if z_0 lies on the noneuclidean segment between z_1 and z_2. In view of the invariance of the distance under noneuclidean motions, we may suppose that $z_0 = 0$ and z_2 is positive real. Then

$$\delta(z_1, z_2) = \log \frac{1+t}{1-t}, \qquad t = \frac{|z_2 - z_1|}{|1 - \bar{z}_1 z_2|},$$

$$\delta(z_1, z_0) = \log \frac{1+|z_1|}{1-|z_1|}, \qquad \delta(z_0, z_2) = \log \frac{1+|z_2|}{1-|z_2|},$$

and

$$\frac{1+t}{1-t} = \frac{(1+t)^2}{1-t^2} = \frac{(|1-\bar{z}_1 z_2| + |z_2 - z_1|)^2}{(1-|z_1|^2)(1-|z_2|^2)}$$

$$\leqslant \frac{(1+|z_1 z_2| + |z_2| + |z_1|)^2}{(1-|z_1|^2)(1-|z_2|^2)} = \frac{(1+|z_1|)(1+|z_2|)}{(1-|z_1|)(1-|z_2|)}.$$

The equality sign holds only if z_1 is a negative real number or 0. This proves the assertion.

We now consider a number of basic problems of noneuclidean geometry. To lay off a noneuclidean segment of prescribed length l at a point z_1 in $\mathfrak{E}$ in a prescribed direction, we employ an appropriate mapping $\mathfrak{A}$ in Ω to carry z_1 to 0 and to align the given direction with the direction of the positive real half axis. Then we lay off at 0 on that half axis a euclidean segment t obtained, in accordance with (6), from the equation

$$l = \log \frac{1+t}{1-t}$$

and equal to

$$t = \frac{e^l - 1}{e^l + 1} = \tanh\left(\frac{l}{2}\right), \tag{8}$$

and apply to its end point the inverse transformation $\mathfrak{A}^{-1}$. This construction can also be used to bisect a noneuclidean segment of length l given by its end points z_1 and z_2, for that problem amounts to laying off a segment of length $l/2$ at z_1 in the direction from z_1 to z_2.

It is natural to define the *noneuclidean circle of radius l about* z_1 as the locus of points z satisfying the equation $\delta(z, z_1) = l$. With $z_1 = 0$ we have $|z| = t$, where t is given by (8). In other words, we are dealing with the euclidean circle of radius t and center 0 orthogonal to the lines through 0. A mapping belonging to the group Ω which carries 0 into a preassigned point z_1 carries euclidean circles and lines into euclidean circles and lines and preserves euclidean angles. It follows that the noneuclidean circles about z_1 belong to the class of euclidean circles orthogonal to the class of noneuclidean lines through z_1; it should be pointed out that for $z_1 \neq 0$ these euclidean circles do not have z_1 as center.

Since the mappings in Ω preserve (the euclidean measure of) angles, it is reasonable to define the noneuclidean measure of a noneuclidean angle as its euclidean measure. On a designated side of a given noneuclidean half line issuing from a point z in $\mathfrak{E}$ there is another noneuclidean half line which forms with the given half line a prescribed angle. (This is obvious for $z = 0$, since the noneuclidean half lines issuing from the center 0 of $\mathfrak{E}$ are its euclidean radii. To go from this special case to the general case we need only make use of an appropriate noneuclidean motion.) In particular, there is precisely one perpendicular to a given noneuclidean line at each of its points. Now we show that, just as in euclidean geometry, the perpendicular bisector of a noneuclidean segment with end points z_1 and z_2 (that is, the noneuclidean line which bisects the segment and is perpendicular to it) is the locus of points equidistant (in the noneuclidean sense) from z_1 and z_2. It suffices to give a proof in the special case when the center z_0 of the segment is 0 and its end points are $z_2 = t > 0$ and $z_1 = -t$. In this case the perpendicular bisector of our segment is the diameter of $\mathfrak{E}$ on the imaginary axis. Let K_1 be the noneuclidean circle of radius l about z_1, and K_2 the image of K_1 under the

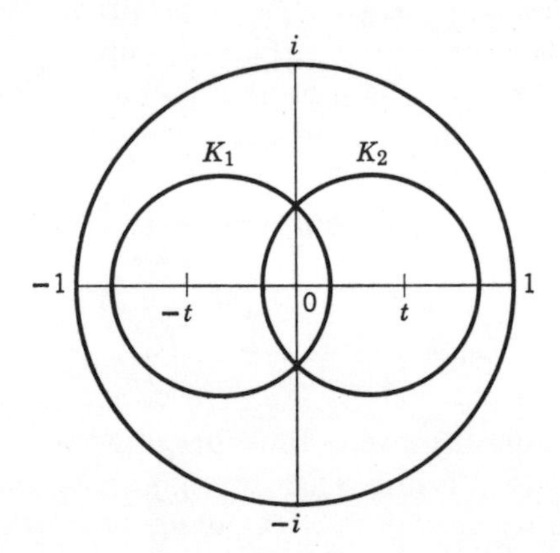

Figure 10

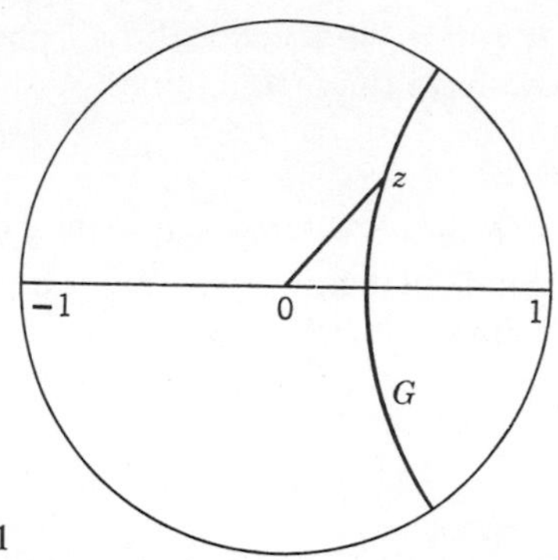

Figure 11

reflection $w = -\bar{z}$ in the imaginary axis (Figure 10). Since this mapping carries z_1 into z_2 and preserves noneuclidean lengths, K_2 is the noneuclidean circle of radius l about z_2. If K_1 and K_2 intersect, then the two points of intersection lie on the imaginary axis. Conversely, if $z = iy$ is a point on the perpendicular bisector, then

$$\delta(z, z_1) = \log\frac{1+q}{1-q}, \qquad q = \frac{|z_1 - z|}{|1 - \bar{z}z_1|} = \frac{|z_2 - z|}{|1 - \bar{z}z_2|}, \qquad \delta(z, z_1) = \delta(z, z_2).$$

This completes the proof.

It is natural to define the perpendicular from a point z_1 in $\mathfrak{E}$ to a noneuclidean line G as the noneuclidean line passing through z_1 and orthogonal to G. To prove existence and uniqueness we may suppose $z_1 = 0$ and G a euclidean arc orthogonal to the positive real half axis. Then the diameters of the unit circle yield the noneuclidean lines through 0. Of these only the one lying on the real axis is orthogonal to G and so yields the required perpendicular (Figure 11). Since of the points z on G the real point has the least absolute value and since the noneuclidean distance

$$\delta(z, 0) = \log\frac{1+|z|}{1-|z|}$$

is a strictly monotonically increasing function of $|z|$, it follows that the shortest of all distances between a point z on G and z_1 is the one measured along the perpendicular. This is another analogy between euclidean and noneuclidean geometry.

Theorem 3: Let z_1, z_2 be two fixed points of $\mathfrak{E}$ and z a variable point of $\mathfrak{E}$, and let us define

$$\gamma(z) = \delta(z, z_2) - \delta(z, z_1).$$

Then $\gamma(z) = 0$ is the equation of the perpendicular bisector M of the noneuclidean segment joining z_1 to z_2. Furthermore, $\gamma(z) > 0$ in the open noneuclidean half plane bounded by M and containing z_1, and $\gamma(z) < 0$ in the open noneuclidean half plane bounded by M and containing z_2. On M we have the inequality

$$2\delta(z, z_1) \geqslant \delta(z_2, z_1).$$

Proof: We have already shown that $\delta(z, z_1) = \delta(z, z_2)$ for z on the perpendicular bisector. If z_0 is again the midpoint of the noneuclidean segment joining z_1 and z_2, then z_0 is the foot of the perpendicular from z_1 to M and, in view of what was proved before, for all z on M we have the inequality

$$\delta(z_2, z_1) = 2\delta(z_0, z_1) \leqslant 2\delta(z, z_1).$$

Now let z_3 be a point of the open noneuclidean half plane $\mathfrak{D}$ containing z_1 and bounded by M. In view of the convexity of $\mathfrak{D}$, the noneuclidean segment V from z_1 to z_3 lies in $\mathfrak{D}$ (Figure 12). Since $\gamma(z)$ is continuous on V and $\gamma(z_1) = \delta(z_1, z_2) > 0$, it follows that $\gamma(z_3) > 0$. Similarly, $\gamma(z) < 0$ on the other side of M. This completes the proof.

We now come to a characteristic difference between euclidean and noneuclidean geometry, namely, the matter of the euclidean parallel axiom. In plane euclidean geometry, through every point off a preassigned line there passes exactly one parallel. This line then does not intersect the given line and has the additional property that all of its points are equidistant from the given line. Now we turn to noneuclidean geometry. Let G be a noneuclidean line not passing through 0. Its ends z_1 and z_2 determine on $|z| = 1$ a great arc B whose length exceeds π. Clearly, every diameter of the unit circle with end points on B forms a noneuclidean line passing through the fixed point 0 and not intersecting G (Figure 13). One of these shares with G the end z_1 and another the end z_2. None of these lines, however, has the property that all of its points are equidistant in the noneuclidean sense from G. To see this, we investigate the locus of all points whose distance from a noneuclidean line G has a preassigned positive value l.

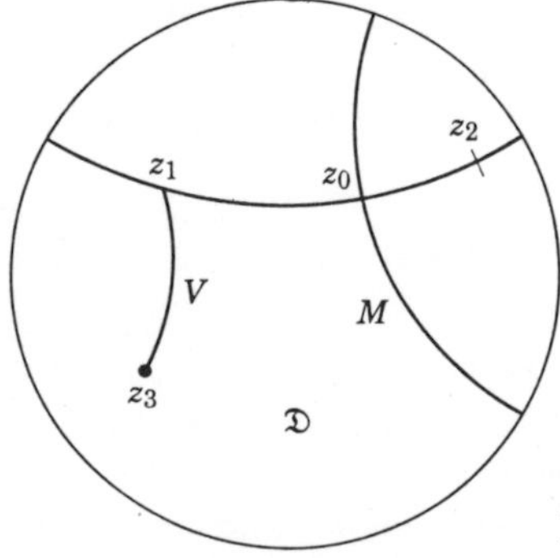

Figure 12

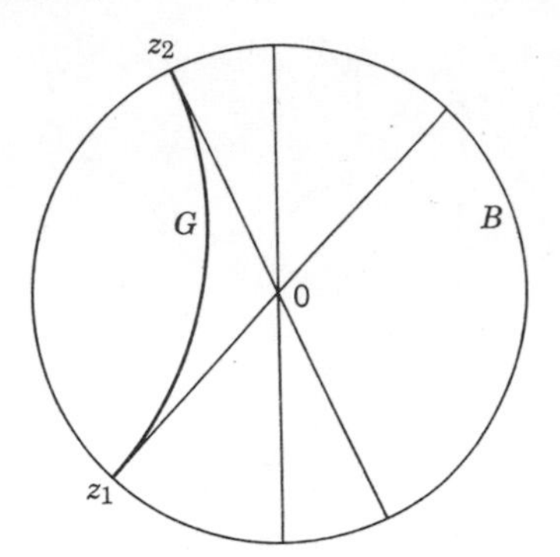

Figure 13

With $\mathfrak{H}$ as our model we can take G to be the positive imaginary half axis. Then the noneuclidean lines orthogonal to G are the semicircles in $\mathfrak{H}$ with center at 0. On one of these semicircles we lay off, on either side of the imaginary axis, a noneuclidean segment of length l and denote the resulting end points by z_1 and z_2. Since the noneuclidean distance in $\mathfrak{H}$ is not affected by a dilation $w = \mu z$ with positive real μ, it follows that the required locus consists of the two euclidean half lines issuing from 0 and passing through z_1 and z_2 (Figure 14). These half lines are not noneuclidean lines, for they are not perpendicular to the real axis. Going back to an arbitrary noneuclidean line G in the model $\mathfrak{E}$ we see that, for a variable l, the equidistant curves are precisely the euclidean circular arcs in $\mathfrak{E}$ passing through the ends of G; of these only G itself is orthogonal to the unit circle (Figure 15).

We now define the noneuclidean arc length for arbitrary curves in $\mathfrak{E}$. We proceed as in the euclidean case. If $z = z(\tau)$ $(a \leqslant \tau \leqslant b)$ is a curve C in $\mathfrak{E}$, then, for all subdivisions $a = \tau_0 < \tau_1 < \cdots < \tau_n = b$ of the curve into finitely many subarcs with $z_k = z(\tau_k)$ $(k = 0, \ldots, n)$, we form the corresponding sums $\delta(z_0, z_1) + \delta(z_1, z_2) + \cdots + \delta(z_{n-1}, z_n)$ and define the *noneuclidean length* s of C as their least upper bound. Using (7) we see that

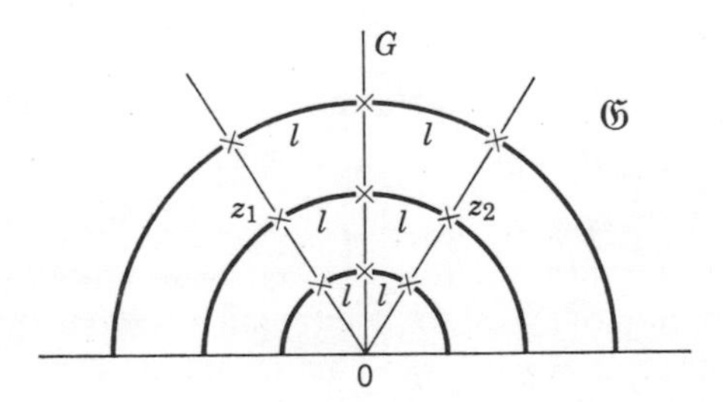

Figure 14

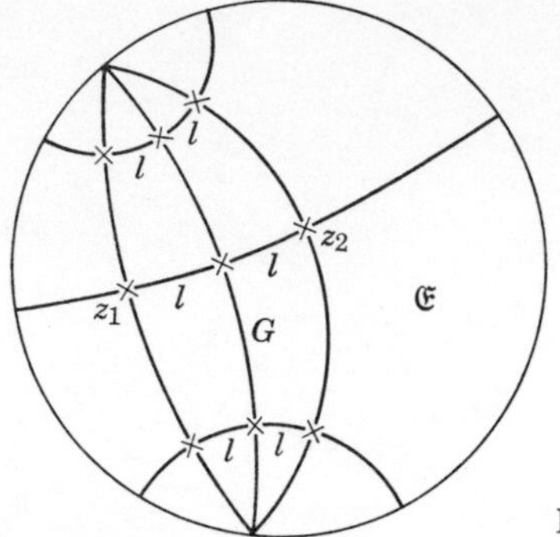

Figure 15

s is finite if and only if C is rectifiable in the euclidean sense, in which case

$$s = \int_C \frac{2\,|dz|}{1 - |z|^2}.$$

This justifies calling the expression

$$ds = \frac{2\,|dz|}{1 - |z|^2}, \tag{9}$$

which is invariant under Ω^*, the *noneuclidean element of arc length.* Just as in euclidean geometry, using the triangle inequality, we can deduce from the definition of s as least upper bound the fact that the noneuclidean segment is the shortest curve joining two arbitrary points z_1, z_2 of $\mathfrak{E}$, and that, for that curve, $s = \delta(z_1, z_2)$. With $\mathfrak{H}$ in place of $\mathfrak{E}$ we must replace z with $(z - i)(z + i)^{-1}$, so that the corresponding expression for the element of arc length is

$$ds = 2(|z + i|^2 - |z - i|^2)^{-1}\,|2i\,dz| = 2i(z - \bar{z})^{-1}\,|dz| = \frac{|dz|}{y} \qquad (z = x + iy).$$

The invariance of the line element ds under Δ^*, that is, under the mappings

$$w = \frac{az + b}{cz + d}, \qquad w = \frac{a\bar{z} - b}{c\bar{z} - d} \qquad (ad - bc = 1), \tag{10}$$

with real a, b, c, d, can be easily checked by direct computation, for

$$|dw| = |cz \pm d|^{-2}\,|dz|, \qquad w - \bar{w} = |cz \pm d|^{-2}(z - \bar{z}).$$

We now show that, conversely, this invariance property determines Δ^* uniquely. For let $u = u(x, y)$, $v = v(x, y)$ be a one-to-one continuously differentiable mapping of $\mathfrak{H}$ onto itself which leaves invariant the noneuclidean line element $ds = y^{-1}\,|dz|$. This means that the equation

$$v^{-2}((du)^2 + (dv)^2) = y^{-2}((dx)^2 + (dy)^2),$$

with $du = u_x\,dx + u_y\,dy$, $dv = v_x\,dx + v_y\,dy$, holds identically in dx and dy. Putting, for brevity, $y^{-1}v = q$, we have

$$u_x^2 + v_x^2 = q^2, \qquad u_x u_y + v_x v_y = 0, \qquad u_y^2 + v_y^2 = q^2$$

so that the 2 by 2 matrix with the elements $q^{-1}u_x$, $q^{-1}u_y$, $q^{-1}v_x$, $q^{-1}v_y$ is orthogonal. It follows that

$$u_x = \varepsilon v_y, \quad u_y = -\varepsilon v_x, \quad u_x v_y - u_y v_x = \varepsilon q^2,$$

with $\varepsilon = \pm 1$. If $\varepsilon = 1$, then $u + iv = f(z)$ is a regular function of z in $\mathfrak{H}$ which is one-to-one in $\mathfrak{H}$ and effects a conformal mapping of $\mathfrak{H}$ onto itself. We know, however, that such a mapping belongs to Δ. If $\varepsilon = -1$, then $u + iv = f(\bar{z})$ is a regular function of $\bar{z} = x - iy$, and it follows, accordingly, that the mapping must have the second form in (10).

We show finally that the group Δ determines the invariant line element uniquely up to a constant factor. If we put $E = y^{-2}$, $F = 0$, $G = y^{-2}$, then the square of the noneuclidean line element is

$$(ds)^2 = E(dx)^2 + 2F(dx)(dy) + G(dy)^2. \tag{11}$$

Suppose the expression on the right side of (11), involving three unknown functions E, F, G in x and y, invariant under all the mappings in Δ. Using the translation $w = z + b$ with arbitrary real b, we see that E, F, G depend on y alone. Using the dilation $w = az$ with arbitrary positive real a, we see that y^2E, y^2F, y^2G are constant, so that

$$y^2(ds)^2 = \lambda(dz)^2 + \mu(dz)(d\bar{z}) + \nu(d\bar{z})^2,$$

with certain constants λ, μ, ν. Finally, using $w = -z^{-1}$, that is, $w - \bar{w} = (z - \bar{z})(z\bar{z})^{-1}$, $dw = z^{-2}\,dz$, $d\bar{w} = \bar{z}^{-2}\,d\bar{z}$, we see that $\lambda = \nu = 0$. Our assertion follows.

The *noneuclidean area element of* $\mathfrak{H}$ is given by

$$d\omega = \sqrt{EG - F^2}\,dx\,dy = y^{-2}\,dx\,dy.$$

The invariance under Δ^* is the consequence of the invariance of the line element ds and can also be easily established directly. We now obtain an expression for $d\omega$ in $\mathfrak{E}$. In view of (9) we have (in $\mathfrak{E}$)

$$E = G = 4(1 - |z|^2)^{-2}, \qquad F = 0, \qquad \sqrt{EG - F^2} = 4(1 - |z|^2)^{-2},$$

and, therefore,

$$d\omega = 4(1 - |z|^2)^{-2}\,dx\,dy. \tag{12}$$

By a *noneuclidean polygon* in $\mathfrak{E}$ we shall mean a subset of $\mathfrak{E}$ bounded by a simple curve consisting of finitely many noneuclidean segments. Let n be the number of vertices of the polygon and let $\alpha_1, \ldots, \alpha_n$ be the interior angles at these vertices.

Theorem 4: The noneuclidean area of a noneuclidean polygon is

$$I = (n - 2)\pi - (\alpha_1 + \cdots + \alpha_n).$$

Proof: For the model $\mathfrak{H}$ we have

$$I = \iint_{\mathfrak{F}} \frac{dx\, dy}{y^2},$$

where $\mathfrak{F}$ denotes the polygonal surface. If $C_1, \ldots, C_n$ are the positively oriented sides of the polygon in their natural order, then, as a special case of Green's formula, we obtain by integration with respect to y the relation

$$I = \sum_{k=1}^{n} \int_{C_k} \frac{dx}{y}, \tag{13}$$

where the individual summands are line integrals. The sides C_k of our polygon are circular arcs in $\mathfrak{H}$ which are orthogonal to the real axis, and it may be assumed that none of the circles involved is a straight line. If r is the radius and a the center of the euclidean circular arc C_k, then we introduce polar coordinates $x - a = r \cos \varphi$, $y = r \sin \varphi$ and obtain

$$\int_{C_k} y^{-1}\, dx = \int_{\beta_k}^{\gamma_k} (r \sin \varphi)^{-1}(-r \sin \varphi)\, d\varphi = -\int_{\beta_k}^{\gamma_k} d\varphi = \beta_k - \gamma_k,$$

where β_k and γ_k are the values of φ at the initial and terminal points of C_k (Figure 16). By (13) we have

$$I = \sum_{k=1}^{n} (\beta_k - \gamma_k).$$

The difference $\gamma_k - \beta_k$ is the change in the direction of the inner normal over C_k. When the inner normal traverses the polygon once, its total rotation is 2π, for the boundary of the polygon is a simple closed curve. Here we

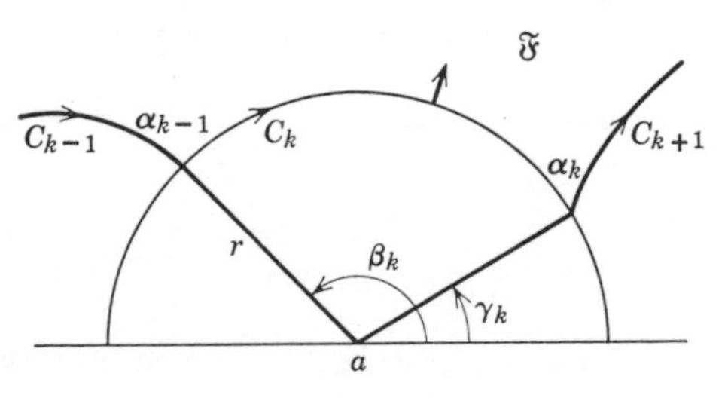

Figure 16

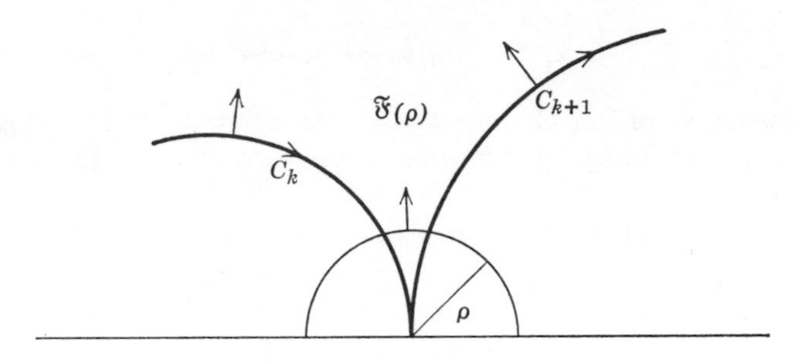

Figure 17

note that at each vertex the angle of the normal changes by an amount equal to the exterior angle associated with the appropriate angle α_k, that is, by $\pi - \alpha_k$. Hence

$$2\pi = \sum_{k=1}^{n}(\gamma_k - \beta_k) + \sum_{k=1}^{n}(\pi - \alpha_k),$$

and our assertion follows.

Theorem 4 remains valid for polygons with improper vertices, that is, vertices lying on the boundary of the unit circle $\mathfrak{E}$. In that case some of the bounding sides of the polygon are noneuclidean half lines or lines whose ends are the vertices in question. The corresponding angles α_k are 0. Also, I is then an improper integral, for the integrand is infinite at the improper vertices. To prove Theorem 4 in this case, we draw about each improper vertex a euclidean circle of sufficiently small radius ρ and disregard the parts of $\mathfrak{F}$ interior to these circles (Figure 17). If the number of improper vertices is m, then we obtain in this way a noneuclidean polygon $\mathfrak{F}(\rho)$ with $m + n$ sides to which one can apply Theorem 4 in the form in which it was proved above. If we note that as $\rho \to 0$, the two vertices of $\mathfrak{F}(\rho)$ near an improper vertex of $\mathfrak{F}$ contribute the amount π to the sum of the angles, then we see that Theorem 4 can indeed be extended to polygons with improper vertices.

Theorem 4 shows that the noneuclidean area of a noneuclidean n-gon is always $\leqslant (n - 2)\pi$, and, in fact, equal to $(n - 2)\pi$ only in the case when all the vertices are improper. In particular, the noneuclidean area of a noneuclidean triangle is equal to the difference of π and the sum of its angles, and so the latter is invariably less than π. This is another characteristic difference between euclidean and noneuclidean geometry.

Theorem 4 can also be proved using the Gauss–Bonnet formula if we note that the noneuclidean lines are the geodesics in the metric (9) and that (as shown by direct calculation) the associated Gaussian curvature has the constant value -1.

3. *Discontinuous groups*

In the remainder of this chapter we deal with discontinuous subgroups of Ω. With a view to future applications we give a rather general definition of this concept.

Let $\mathfrak{G}$ be a space of points $\mathfrak{p}$ for which neighborhoods are defined in the usual way. We call a group Γ of single-valued mappings $A, B, \ldots$ of $\mathfrak{G}$ onto itself *discontinuous* in $\mathfrak{G}$ if there is no point in $\mathfrak{G}$ all of whose images under the mappings in Γ have a limit point in $\mathfrak{G}$. Note that, for Γ discontinuous in $\mathfrak{G}$, there are only finitely many elements of Γ which map an arbitrary point of $\mathfrak{G}$ onto itself. Since they form a group, our mappings must be one-to-one. Furthermore, it is clear that in the case of a compact space $\mathfrak{G}$ every discontinuous group must be finite.

An important simple example of a discontinuous group is the group of translations of the finite z-plane

$$(1) \qquad w = z + \omega, \qquad \omega = k\omega_1 + l\omega_2 \qquad (k, l = 0, \pm 1, \pm 2, \ldots),$$

where ω_1 and ω_2 are two distinct nonzero numbers with nonreal ratio. This group appeared in Section 9 of Chapter 1 (Volume I). On the other hand, the rotations

$$w = \lambda^n z \qquad (|\lambda| = 1; \ n = 0, \pm 1, \ldots),$$

where $\lambda = e^{\pi i a}$ and a is real and irrational, form a group of mappings of the circle $|z| = 1$ onto itself which is not discontinuous. Here $\mathfrak{G}$ is compact, while Γ has infinite order because we assumed that a is irrational.

A discontinuous group Γ, then, is linked to a space which is mapped onto itself by the elements of Γ. The case when the space $\mathfrak{G}$ is itself a group Ω with a system of neighborhoods, namely, the case of a *continuous* or *topological group*, is of particular importance. The special type of continuous group with which we shall be concerned has the following additional property. The elements X of the group Ω depend on finitely many real parameters, and we can, therefore, introduce n independent real local coordinates $x_1, \ldots, x_n$. The coordinates of the product XY are continuous functions of the local coordinates of X and Y. A corresponding statement holds for the coordinates of the inverse X^{-1}. In the sequel the term "continuous group" will always refer to a continuous group with this additional property. Now let Γ be a subgroup of the continuous group Ω. If A is a given element of Γ, then $Y = AX$, with X varying over Ω, defines a one-to-one mapping of the continuous group Ω onto itself. It is clear that different elements A and B of Γ define different mappings of Ω, and, since the relations $Y = AX$, $X = BZ$, $AB = C$ imply the relation $Y = CZ$, these mappings are composed in the manner of the associated group elements. This enables us to view Γ as a group of mappings of the continuous group Ω onto itself. If Γ is discontinuous in the topological group Ω, then Γ is called a *discrete subgroup* of Ω.

An example of a continuous group is furnished by the 2 by 2 matrices

$$X = \begin{pmatrix} x_1 & x_2 \\ x_3 & x_4 \end{pmatrix},$$

with x_1, x_2, x_3, x_4 real and $x_1x_4 - x_2x_3 \neq 0$. This topological group Ω can be thought of as arising from four-dimensional euclidean space by omission of the cone $x_1x_4 = x_2x_3$. The X with x_1, x_2, x_3, x_4 integers and $x_1x_4 - x_2x_3 = \pm 1$ form a discrete subgroup Γ.

Theorem 1: A subgroup Γ of a continuous group Ω is discrete if and only if a sufficiently small neighborhood of the unit element E of Ω contains no elements of Γ other than E.

Proof: We suppose Γ not discrete, that is, not discontinuous in the topological group Ω. Then there is an element X_0 of Ω such that, for A varying in Γ, the elements AX_0 have a limit point Y_0 in Ω. Now let A_k $(k = 1, 2, \ldots)$ be a sequence of distinct elements of Γ for which $A_kX_0 \to Y_0$ $(k \to \infty)$. Since Ω is a continuous group, we have, in particular, $A_{k+1}A_k^{-1} \to E$ $(k \to \infty)$, so that every neighborhood of E contains an element of Γ distinct from E. Conversely, if $B_k \neq E$ $(k = 1, 2, \ldots)$ is a sequence of distinct elements of Γ converging to E, then $B_kX \to X$ for every X in Ω and every point of the topological group is a limit point of its images AX for every mapping A in Γ. This completes the proof.

Throughout the remainder of this chapter, Ω denotes the multiplicative group of 2 by 2 matrices

$$M = \begin{pmatrix} p & \bar{q} \\ q & \bar{p} \end{pmatrix} \tag{2}$$

with complex p, q and $p\bar{p} - q\bar{q} = 1$. If we put $p = y_1 + y_2i$, $q = y_3 + y_4i$, then we obtain a representation of our topological group by means of the three-dimensional surface $y_1^2 + y_2^2 - y_3^2 - y_4^2 = 1$. Every element M of Ω yields a fractional linear transformation

$$w = \frac{\bar{p}z + \bar{q}}{qz + p} \tag{3}$$

of the unit circle $\mathfrak{E}$ onto itself. Here it should be noted that M and $-M$ define the same transformation (3) although they are distinct elements of Ω. Previously (Section 2) Ω denoted the group of fractional linear transformations (3). With the new meaning of Ω, the group of transformations (3) is isomorphic to the factor group of Ω with respect to the invariant subgroup of order two consisting of E and $-E$. We obtain an example of a discrete subgroup of Ω if we take as y_1, y_2, y_3, y_4 in $p = y_1 + y_2 i$, $q = y_3 + y_4 i$ the various integral solutions of the diophantine equation $y_1^2 + y_2^2 - y_3^2 - y_4^2 = 1$. In fact, since

$$M^{-1} = \begin{pmatrix} p & -\bar{q} \\ -q & \bar{p} \end{pmatrix},$$

we see that the matrices in question form a group Γ which is discrete because the integers have no limit point. Since $y_1 = 2m + 1$, $y_2 = m - 1$, $y_3 = 2m$, $y_4 = m + 1$ is a solution for every m, it follows that Γ has infinitely many elements.

We now come to an important but simple theorem which characterizes discrete subgroups of Ω.

Theorem 2: A subgroup of Ω is discrete if and only if it is discontinuous in $\mathfrak{E}$.

Proof: Let Γ be a subgroup of Ω which is not discrete. By Theorem 1, Γ contains a sequence of matrices (2) with $M \to E$, $M \neq E$, that is, with $p \to 1$, $q \to 0$ (here we may assume that not both M and $-M$ appear in the sequence). The corresponding distinct transformations (3) give a sequence of images $w = \bar{q}p^{-1}$ of the point $z = 0$ which converge to 0. This shows that Γ is not discontinuous in $\mathfrak{E}$.

Conversely, assume Γ is not discontinuous in $\mathfrak{E}$. Then there is a point z_0 in $\mathfrak{E}$ and a sequence of distinct elements

$$M_k = \begin{pmatrix} \bar{p}_k & \bar{q}_k \\ q_k & p_k \end{pmatrix} \qquad (k = 1, 2, \dots)$$

of Γ such that the images w_k of z_0 under the corresponding transformations (3) converge to a point ζ in $\mathfrak{E}$. By (14), Section 1,

$$\frac{1 - |z_0|^2}{|p_k + q_k z_0|^2} = 1 - |w_k|^2 \to 1 - |\zeta|^2 > 0,$$

$$|p_k + q_k z_0|^2 \to \frac{1 - |z_0|^2}{1 - |\zeta|^2} \qquad (k \to \infty). \tag{4}$$

On the other hand, $|p_k|^2 - |q_k|^2 = 1$ implies

$$\left|\frac{q_k}{p_k}\right| < 1, \qquad |p_k + q_k z_0| = |p_k| \left| 1 + \frac{q_k}{p_k} z_0 \right| \geqslant |p_k| (1 - |z_0|).$$

Since the factor $1 - |z_0|$ above is a positive number independent of k, (4) implies that the p_k are bounded and, since $|q_k| < |p_k|$, the q_k are also bounded.

It follows that the sequences of numbers p_k and q_k $(k = 1, 2, \ldots)$ have finite limit points p and q. But then the matrices M_k $(k = 1, 2, \ldots)$ have a limit point M in Ω. This shows that Γ is not discrete and the proof is complete.

To put the assertion of the theorem in the proper light, we consider in place of the open unit disk $\mathfrak{E}$ its boundary $|z| = 1$. This set is also mapped onto itself by all the elements M of Ω. Since it is compact we see that a subgroup Γ of Ω is discontinuous on the boundary of the unit disk only if it is finite.

We know from Section 10 of Chapter 2 (Volume I) that the subgroups of Ω which are discontinuous in $\mathfrak{E}$ play an important role in the theory of functions. No procedure is known, however, for obtaining all of these groups. One class of discontinuous groups Γ was given by Jakob Nielsen whose result appears below as Theorem 3. Let us call a subgroup Γ of Ω *hyperbolic* if all its elements other than $\pm E$ are hyperbolic. In view of Theorem 5, Section 1, this means that all the elements $M \neq \pm E$ of Γ satisfy the condition $(p + \bar{p})^2 > 4$, so that the fixed points of the corresponding transformations (3) must lie on the boundary of the unit disk. An example of such a group Γ is given by all matrices

$$M = \begin{pmatrix} p & q \\ q & p \end{pmatrix} \qquad (p^2 - q^2 = 1), \tag{5}$$

with p and q real numbers. Here the fixed points are 1 and -1. Nielsen's result deals with hyperbolic subgroups of Ω; specifically, it asserts that

Theorem 3: Every hyperbolic noncommutative subgroup of Ω is discontinuous in $\mathfrak{E}$.

Proof: By assumption, the given group Γ is nonabelian. Hence there is some pair of elements A, B in Γ with $AB \neq BA$; in particular $A \neq \pm E$. Instead of $\mathfrak{E}$ we work with the upper half plane $\mathfrak{H}$. Then Ω and Γ are replaced by $C^{-1}\Omega C = \Delta$ and $C^{-1}\Gamma C = \Gamma_0$, where C is the matrix of the transformation $w = (z - i)(z + i)^{-1}$. The elements of Δ are given by

$$B = \begin{pmatrix} a & b \\ c & d \end{pmatrix} \qquad (ad - bc = 1),$$

with a, b, c, d real numbers. The element B is hyperbolic if and only if $(a + d)^2 > 4$ and, in that case, the fixed points of the transformation

$$w = \frac{az + b}{cz + d}$$

lie on the real axis. Now, let A and B be two elements of Γ_0 with $A \neq \pm E$, that is, A hyperbolic. Then the commutator $G = ABA^{-1}B^{-1}$ is either $\pm E$

or hyperbolic. We may assume, with no loss of generality, that the two fixed points of A are 0 and ∞ so that

$$A = \begin{pmatrix} \rho & 0 \\ 0 & \rho^{-1} \end{pmatrix}, \tag{6}$$

with $\rho \neq \pm 1$ a real number. We have

$$G = \begin{pmatrix} \rho & 0 \\ 0 & \rho^{-1} \end{pmatrix}\begin{pmatrix} a & b \\ c & d \end{pmatrix}\begin{pmatrix} \rho^{-1} & 0 \\ 0 & \rho \end{pmatrix}\begin{pmatrix} d & -b \\ -c & a \end{pmatrix} = \begin{pmatrix} ad - bc\rho^2 & ab(\rho^2 - 1) \\ cd(\rho^{-2} - 1) & ad - bc\rho^{-2} \end{pmatrix}$$

$$G = E - (\rho - \rho^{-1})\begin{pmatrix} bc\rho & -ab\rho \\ cd\rho^{-1} & -bc\rho^{-1} \end{pmatrix}. \tag{7}$$

It follows that the trace of the matrix G has the real value

$$\sigma(G) = 2 - bc(\rho - \rho^{-1})^2, \tag{8}$$

whose square is at least 4. This implies the inequality

$$(bc)^2(\rho - \rho^{-1}) \geqslant 4bc. \tag{9}$$

Furthermore, the commutator $AGA^{-1}G^{-1} = H$ of A and G is also either $\pm E$ or hyperbolic, so that, by (7) and (8), the square of

$$\sigma(H) = 2 + abcd(\rho - \rho^{-1})^4$$

is at least 4. This means that

$$(abcd)^2(\rho - \rho^{-1})^4 \geqslant -4abcd. \tag{10}$$

We now assume that the subgroup Γ is not discontinuous in $\mathfrak{E}$. Then, by Theorem 2, it is not discrete and, by Theorem 1, we can find in Γ_0 a sequence of matrices $B \neq E$ converging to E. With A in (6) fixed, ρ is fixed and $(\rho - \rho^{-1})^2$ is a fixed positive number. Since $B \to E$, the products $bc \to 0$, so that, by (9),

$$bc \leqslant 0 \tag{11}$$

for almost all terms B of our sequence. On the other hand, the products $ad \to 1$, so that, for almost all B, $adbc$ has the same sign as bc. Since $adbc \to 0$, (10) shows that

$$bc \geqslant 0 \tag{12}$$

for almost all terms of our sequence. Together, (11) and (12) imply that $bc = 0$ for almost all B, so that, by (8), $\sigma(G) = 2$. Since G is either $\pm E$ or hyperbolic, we can now conclude that $G = E$. But then, by (7), $ab = 0$ and $cd = 0$. Since $ad \to 1$, we find $b = 0$ and $c = 0$ for almost all B, which must, as a result, have the same diagonal form as A. This means that there is a sequence of matrices in Γ_0 converging to E. We denote the general term of this sequence by A, where A is of the form (6), except that now instead of being fixed ρ runs through a sequence of values $\neq 1$ which converge to 1. Now let B denote a fixed but arbitrary element of Γ_0. Then, as $A \rightarrow E$, the commutators G of A and B also tend to E. Applying to this sequence of $G \rightarrow E$ the results established above for the sequence of $B \rightarrow E$, we conclude that almost all the G must be in diagonal form. Then (7) implies $ab = 0$ and $cd = 0$. The condition $a = 0$ would imply $-bc = ad - bc = 1$ and so $c \neq 0$, $d = 0$. But then the trace of B would be 0 and, contrary to the assumption about Γ, B would be elliptic. Therefore, $a \neq 0$, $b = 0$, $ad = ad - bc = 1$, $d \neq 0$, $c = 0$. It follows that every element B of Γ_0 is in diagonal form. Since this shows that Γ_0 is a commutative group, the original assumption that Γ is not discontinuous must be false. This completes the proof of Theorem 3.

It should be noted that the group given by the matrices (5) is not discontinuous in $\mathfrak{E}$, but, in accordance with Theorem 3, it is commutative. It is easy to show that a commutative hyperbolic group Γ is discontinuous in $\mathfrak{E}$ if and only if every mapping in Γ is a power of a basis element, that is, Γ is cyclic. We do not prove this assertion because we shall not make use of it in the sequel.

There is a connection between the result above and the investigations of Chapter 2. As we shall see later, the uniformization of algebraic function fields of genus $p > 1$ leads to noncommutative hyperbolic subgroups of Ω. At this point, however, we pursue our study of arbitrary subgroups of Ω which are discontinuous in $\mathfrak{E}$, and call them, for brevity, *circle groups*. Next we introduce the concept of a fundamental region in a form sufficiently general to meet the requirements of future applications. Let $\mathfrak{G}$ be a region in a finite-dimensional euclidean space and let Γ be a discontinuous group in $\mathfrak{G}$. A set $\mathfrak{F}$ in $\mathfrak{G}$ is called a *fundamental region* of Γ if it has euclidean content in the Jordan sense and if all the images $\mathfrak{F}_A$ of $\mathfrak{F}$, A in Γ, cover $\mathfrak{G}$ simply and without gaps. More precisely, the $\mathfrak{F}_A$ cover $\mathfrak{G}$ completely and have no interior points in common. We note that $\mathfrak{F}$ need not be connected. In applications $\mathfrak{F}$ will always arise from a domain by adjunction of its boundary points. One example is the period parallelogram of the group of translations (1) with ω_1 and ω_2 the fundamental periods of an elliptic function (cf. Section 10, Chapter 1, Volume I).

Throughout the remainder of this chapter $\mathfrak{G} = \mathfrak{E}$ and Γ is a circle group. Note that if $-E$ is an element of Γ, then A in Γ implies $-A$ in Γ. From now on in all such cases Γ shall denote the factor group resulting from the identification of E and $-E$.

Theorem 4: Every circle group has a fundamental region.

Proof: We shall give a constructive procedure for the determination of a fundamental region with a reasonable boundary. This will be done in three steps.

I. A point of $\mathfrak{E}$ is called a *fixed point* of the given circle group Γ if it is left fixed by at least one mapping in Γ other than the identity. We show first that Γ has at most countably many fixed points. Since an element of Γ has at most one fixed point in $\mathfrak{E}$, it suffices to show that Γ itself has at most countably many elements. If n is a natural number, then the number of elements M in Γ with $|p| \leqslant n$, $|q| \leqslant n$, is finite, for otherwise these M would have a limit point contradicting Theorem 2. Choosing $n = 1, 2, \ldots$, we see readily that Γ is finite or countably infinite.

Let ζ be a point of $\mathfrak{E}$ which is not a fixed point of Γ. If ζ_A is the image of ζ under the mapping A in Γ, then for two distinct mappings A and B, we always have $\zeta_{B^{1'}A} \neq \zeta$, so that $\zeta_A = (\zeta_{B^{1'}A})_B \neq \zeta_B$. It follows that all the images ζ_A are distinct. If Γ has infinitely many elements, then the points ζ_A must have a limit point in the closed disk $|z| \leqslant 1$. Since Γ is discontinuous in $\mathfrak{E}$, it follows that the limit points of the ζ_A lie on the boundary $|z| = 1$ of $\mathfrak{E}$.

Using the point ζ and all of its images ζ_A, we construct a fundamental region of Γ in the following well-defined manner. We consider the noneuclidean segment with end points ζ and ζ_A, $A \neq \pm E$, and construct its perpendicular bisector $L(A)$. This perpendicular bisector separates the noneuclidean plane $\mathfrak{E}$ into two half planes with common boundary $L(A)$. Whenever this boundary line is to be omitted, we refer specifically to an open half plane. Of the two noneuclidean half planes, we consider the one containing the point ζ, and denote it by $\mathfrak{D}(A)$. We propose to show that if A varies over the mappings in Γ distinct from the identity, then the intersection $\mathfrak{F}$ of all the $\mathfrak{D}(A)$ forms a fundamental region of Γ.

II. If the group Γ is finite, then there are only finitely many perpendicular bisectors $L(A)$, and $\mathfrak{F}$ is readily constructed. To get something of an overview of the construction of $\mathfrak{F}$ for an infinite group Γ, we consider a disk $|z| < r (0 < r < 1)$ concentric with and smaller than $\mathfrak{E}$ and show that the number of perpendicular bisectors $L(A)$ entering this disk is finite. Observe that the limit points of the ζ_A lie on the rim of the unit disk, so that the noneuclidean distance $\delta(\zeta, \zeta_A)$ tends to infinity as A varies over Γ. Now, Theorem 3 of Section 2 shows that for every point z on $L(A)$ the following inequality holds:

$$2\delta(z, \zeta) \geqslant \delta(\zeta, \zeta_A). \tag{13}$$

Since for a fixed $r < 1$ the points of the disk $|z| < r$ are a bounded noneuclidean distance away from ζ, it follows that the number of perpendicular bisectors $L(A)$ entering it is indeed finite.

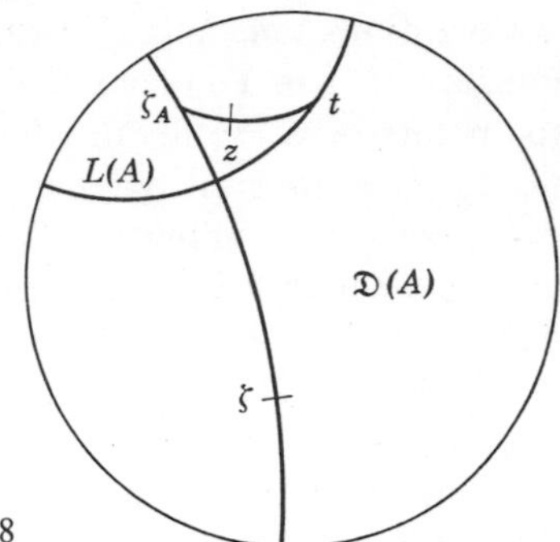

Figure 18

Every half plane $\mathfrak{D}(A)$ is noneuclidean convex and closed in $\mathfrak{E}$. Furthermore, the intersection of a finite or infinite number of convex and closed regions is convex and closed. It follows that $\mathfrak{F}$ is noneuclidean convex and closed in $\mathfrak{E}$. Now we consider the function

$$\delta(\zeta_A, z) - \delta(\zeta, z) = \gamma(A, z) = \gamma(z),$$

with z varying in $\mathfrak{E}$. By Theorem 3 of Section 2, $\gamma(z) \geqslant 0$ if and only if z lies in the noneuclidean half plane $\mathfrak{D}(A)$, and $\gamma(z) = 0$ only when z is on the boundary $L(A)$. This implies that a point z of $\mathfrak{E}$ belongs to $\mathfrak{F}$ if and only if the inequality

$$\delta(\zeta, z) \leqslant \delta(\zeta_A, z) \tag{14}$$

holds for all $\zeta_A \neq \zeta$.

Now we show that z is an interior point of $\mathfrak{F}$ if and only if the strict inequality

$$\delta(\zeta, z) < \delta(\zeta_A, z) \tag{15}$$

holds for all $\zeta_A \neq \zeta$ (Figure 18). In fact, if $\delta(\zeta, z) = \delta(\zeta_A, z)$ for a point $z = t$ of $\mathfrak{F}$ and an image point $\zeta_A \neq \zeta$, then, in view of Theorem 3 of the preceding section, we have $\gamma(z) < 0$ for all points z other than t belonging to the noneuclidean segment with end points ζ_A and t. This means that such a z does not belong to $\mathfrak{F}$, and, as $z \to t$, t is seen to be a boundary point of $\mathfrak{F}$. On the other hand, if (15) holds for all $\zeta_A \neq \zeta$ at a given point $\eta = z$ in $\mathfrak{E}$, then, in any event, η and ζ lie in a disk $|z| < r < 1$ which intersects only finitely many $L(A)$. In other words, for almost all A this disk lies entirely in the half plane $\mathfrak{D}(A)$; so that, for these A, $\gamma(z) > 0$, provided that z belongs to a sufficiently small neighborhood of η independent of A. For the remaining finitely many A we use the continuity of $\gamma(A, z)$ and see that a full neighborhood of η belongs to $\mathfrak{F}$. This shows that η is an interior point of $\mathfrak{F}$.

Conditions (14) and (15) can be formulated differently. If we define the images z_A, A in Γ, of an arbitrary point z of $\mathfrak{E}$ to be *equivalent* to z, then the points of $\mathfrak{E}$ are divided into classes. Since, for variable A and fixed z, the z_A have no limit point in $\mathfrak{E}$, it follows that the noneuclidean distances $\delta(\zeta, z_A)$ have a minimum. We claim that z is a point of $\mathfrak{F}$ if and only if, among all points z_A equivalent to it, z has minimal noneuclidean distance from ζ. In fact, the invariance of noneuclidean distance under the mappings A implies that $\delta(\zeta_A, z) = \delta(\zeta, z_{A^{-1}})$. It therefore follows from (14) that z is in $\mathfrak{F}$ if and only if $\delta(\zeta, z) \leqslant \delta(\zeta, z_{A^{-1}})$ for all A in Γ. Here we may replace A^{-1} with A and obtain the desired assertion that $\mathfrak{F}$ consists of all z satisfying the conditions

$$(16) \qquad \delta(\zeta, z) \leqslant \delta(\zeta, z_A) \qquad (A \text{ in } \Gamma).$$

The strict inequality $\delta(\zeta, z) < \delta(\zeta, z_A)$ holds for all mappings A distinct from the identity if and only if z is an interior point of $\mathfrak{F}$.

III. It is now easy to conclude the proof. First we show that the images $\mathfrak{F}_A$ of $\mathfrak{F}$ cover the unit disk $\mathfrak{E}$. In fact, if z is a point of $\mathfrak{E}$, then some point z_B equivalent to z is a minimal distance away from ζ, and we have

$$\delta(\zeta, z_B) \leqslant \delta(\zeta, z_A)$$

for all A in Γ. If we replace A by AB in this inequality, then (16) is satisfied with z_B in place of z. Hence z_B is in $\mathfrak{F}$ and z is in $\mathfrak{F}_{B^{-1}}$.

Next we show that if A and B are distinct mappings in Γ, then $\mathfrak{F}_A$ and $\mathfrak{F}_B$ cannot have a common interior point. Suppose z were an interior point of $\mathfrak{F}_A$ and a point of $\mathfrak{F}_B$. Then $z_{A^{-1}}$ would be an interior point of $\mathfrak{F}$ and $z_{B^{-1}}$ a point of $\mathfrak{F}$, so that, by (14) and (15), the inequalities

$$\delta(\zeta, z_{A^{-1}}) < \delta(\zeta_C, z_{A^{-1}}), \quad \delta(\zeta, z_{B^{-1}}) \leqslant \delta(\zeta_D, z_{B^{-1}})$$

would hold for all mappings C and D in Γ distinct from the identity. In particular, if we choose $C = A^{-1}B$, $D = B^{-1}A$ and bear in mind the invariance of noneuclidean distance, then, in contradiction to the first inequality, the second inequality implies

$$\delta(\zeta_C, z_{A^{-1}}) \leqslant \delta(\zeta, z_{A^{-1}}).$$

It remains to show that $\mathfrak{F}$ has euclidean content in the Jordan sense. We showed that every boundary point of $\mathfrak{F}$ must lie on at least one of the euclidean arcs $L(A)$ almost all of which lie in an arbitrarily narrow ring $r \leqslant |z| < 1$. It follows that the boundary of $\mathfrak{F}$ can be covered by finitely many squares of arbitrarily small total euclidean content. This completes the proof.

The fundamental region $\mathfrak{F}$ is completely determined by Γ and ζ. We shall refer to it as the *normal fundamental region of* Γ *with center* ζ.

We observe that, in view of (12), Section 2, the noneuclidean area of $\mathfrak{F}$ is given by

$$I(\mathfrak{F}) = 4 \iint_{\mathfrak{F}} \frac{dx\, dy}{(1 - x^2 - y^2)^2}.$$

If $\mathfrak{F}$ is contained in a circle concentric with and smaller than $\mathfrak{E}$, then this is an ordinary, finite double integral. However, if there are boundary points of $\mathfrak{F}$ on the unit circle $|z| = 1$, then $I(\mathfrak{F})$ denotes an improper integral, and its value may be infinite.

4. *Polygon groups*

The circle groups of greatest importance for function-theoretical applications are those for which there exists a fundamental region $\mathfrak{F}$ with finite noneuclidean area $I(\mathfrak{F})$. The following theorems shed light on such groups.

Theorem 1: A normal fundamental region of a circle group has finite noneuclidean area if and only if it is a noneuclidean polygon.

Proof: Let $\mathfrak{F}$ be a normal fundamental region of a circle group Γ. If $\mathfrak{F}$ is a noneuclidean polygon, with or without noneuclidean vertices, then, by Theorem 4, Section 2, $I(\mathfrak{F})$ is finite. It remains to prove the converse. This is done in five steps.

I. Let $\mathfrak{F}$ be a normal fundamental region with $I(\mathfrak{F})$ finite. Since $I(\mathfrak{E})$ is infinite, Γ contains elements other than the identity and $\mathfrak{F}$ has boundary points in $\mathfrak{E}$. If t is such a boundary point, then it lies on at least one perpendicular bisector $L(A)$. For every positive ρ, the inequality $\delta(z, t) < \rho$ defines a neighborhood $\mathfrak{U}$ of t which is the interior of the noneuclidean circle of radius ρ about t. The number ρ can be chosen so small that all the perpendicular bisectors $L(A)$ entering $\mathfrak{U}$ pass through t. We suppose this to be the case for the distinct mappings $A_1, \ldots, A_r$. Since the images ζ_A of the center ζ of $\mathfrak{F}$ are all different and ζ_A is uniquely determined by the perpendicular from ζ to $L(A)$, the r perpendicular bisectors $L(A_k)$ are distinct noneuclidean lines which divide $\mathfrak{U}$ into $2r$ sectors. The noneuclidean segment from t to ζ crosses exactly one of these sectors, which we denote by $\mathfrak{S}$. $\mathfrak{S}$ is precisely the intersection of $\mathfrak{F}$ and $\mathfrak{U}$. In the case $r > 1$, there are two noneuclidean segments belonging to the boundary of $\mathfrak{F}$ which issue from t and enclose the angular segment $\mathfrak{S}$. Then t is a vertex of $\mathfrak{F}$. In the case $r = 1$, t is an interior point of a noneuclidean segment on $L(A)$ which belongs entirely to the boundary of $\mathfrak{F}$. Extension of this segment to one of the two sides leads to a vertex of $\mathfrak{F}$ or to an end of $L(A)$ on $|z| = 1$. A corresponding statement holds when t is a vertex of $\mathfrak{F}$ to begin with, and we extend one of the two noneuclidean boundary segments of $\mathfrak{F}$ issuing from t. It follows that if, beginning at t, we traverse this portion of the boundary of $\mathfrak{F}$ in a definite direction we encounter: (i) a finite number of vertices of $\mathfrak{F}$ and then the rim of the unit disk, or (ii) an infinite sequence of vertices, or (iii) finitely many vertices and then the starting point t.

We call the connected portion of the boundary of $\mathfrak{F}$ on both sides of t the *boundary component R determined by t*. Let $\ldots, z_{-1}, z_0, z_1, \ldots$ denote the sequence, finite or infinite to either side, of vertices of $\mathfrak{F}$ associated with R traversed in the positive direction. In the third case R is a simple closed noneuclidean polygon with ordinary vertices which, in view of the convexity of $\mathfrak{F}$, encloses all of the fundamental region $\mathfrak{F}$; then $\mathfrak{F}$ is surely compact. From now on we can exclude the third case in our proof. In the first case we designate the point of R on $|z| = 1$ as an improper vertex of $\mathfrak{F}$ and include it in the count of vertices. It is conceivable that some boundary points of $\mathfrak{F}$ in $\mathfrak{E}$ lie on boundary components other than that determined by t. However, since the number of perpendicular bisectors $L(A)$ is at most countably infinite, there can be at most countably many boundary components.

II. Now we join the vertices z_k $(k = 0, \pm 1, \ldots)$ on R to the center ζ of $\mathfrak{F}$ by means of noneuclidean segments or half lines. If z_k, z_{k+1} are two successive vertices of $\mathfrak{F}$, then the three points z_k, z_{k+1}, ζ are the vertices of a noneuclidean triangle $\mathfrak{D}_k$ with angles α_k, β_k, γ_k (Figure 19) which, in view of the convexity of $\mathfrak{F}$, is entirely contained in the fundamental region. If z_k or z_{k+1} is an improper vertex, then α_k or β_k is 0. By Theorem 4, Section 2, the noneuclidean area of the triangle $\mathfrak{D}_k$ is

$$I(\mathfrak{D}_k) = \pi - \alpha_k - \beta_k - \gamma_k \qquad (k = 0, \pm 1, \ldots).$$

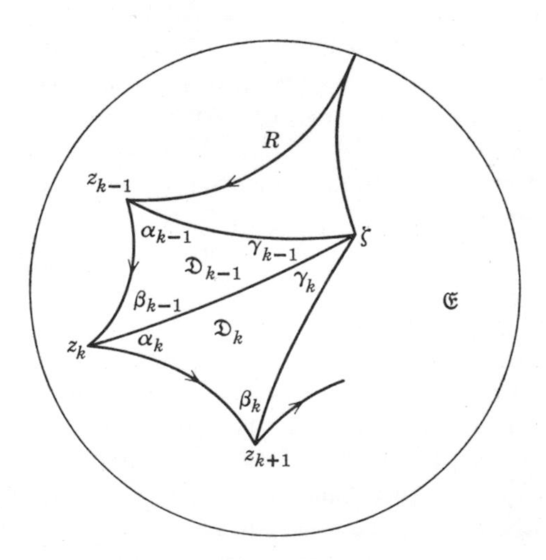

Figure 19

Summation over a connected portion of R yields

(1) $$\sum_{k=m}^{n} I(\mathfrak{D}_k) = \sum_{k=m}^{n} (\pi - \alpha_k - \beta_k - \gamma_k) \qquad (m \leqslant n).$$

If z_k is a proper vertex of $\mathfrak{F}$, then $\alpha_k + \beta_{k-1} = \omega_k$ is the corresponding vertex angle of $\mathfrak{F}$ and, in view of the convexity of $\mathfrak{F}$, we have

(2) $$0 < \omega_k = \alpha_k + \beta_{k-1} < \pi.$$

Upon introduction of the ω_k, (1) becomes

(3) $$\sum_{k=m}^{n} \gamma_k + \sum_{k=m}^{n} I(\mathfrak{D}_k) = \pi - \alpha_m - \beta_n + \sum_{k=m+1}^{n} (\pi - \omega_k) \qquad (m \leqslant n).$$

Now we choose n as large as possible in case (i) and let n go to ∞ in case (ii). Similarly, we choose m as small as possible in case (i) and let m go to $-\infty$ in case (ii). Since all the triangles $\mathfrak{D}_k$ belong to $\mathfrak{F}$ and do not overlap,

(4) $$\sum_{k=m}^{n} I(\mathfrak{D}_k) \leqslant I(\mathfrak{F}).$$

Because the summands on the left are all positive, each of the infinite sums associated with the transitions to the limit converges, and the same is true of

(5) $$\sum_{k=m}^{n} \gamma_k \leqslant 2\pi.$$

By (2), $\alpha_m < \pi$, $\beta_n < \pi$, so that the α_m are bounded for $m \to -\infty$ and the β_n are bounded for $n \to \infty$. Again by (2), the terms $\pi - \omega_k$ of the sum on the right-hand side of (3) are all positive. This fact and the boundedness of its partial sums (implied by (4) and (5)) assure its convergence for $m \to -\infty$ and $n \to \infty$. It now follows that for $m \to -\infty$ the α_m approach a limit $\alpha_{-\infty}$ and that for $n \to \infty$ the β_n approach a limit β_∞. Here we do not claim that either of the sequences z_m or z_n converges. On the other hand, if there is a minimal m or maximal n, then $\alpha_m = 0$ or $\beta_n = 0$.

III. Now we investigate further the case when there is no maximal n and prove that, in that case,

(6) $$\beta_\infty \leqslant \frac{\pi}{2}.$$

Let $r_n = \delta(\zeta, z_n)$. If z_n is on $L(A)$ (Figure 20), then, by (13), Section 3, we have the inequality

$$2r_n \geqslant \delta(\zeta, \zeta_A).$$

Since the ζ_A have no limit point in $\mathfrak{E}$, $|\zeta_A| \to 1$ for $n \to \infty$ and, as a result, $r_n \to \infty$. This means that $r_{n+1} > r_n$ for infinitely many natural numbers n. We now show that in a noneuclidean triangle, too, the larger angle is opposite the larger side, so that $\alpha_n > \beta_n$. To this end, we construct the

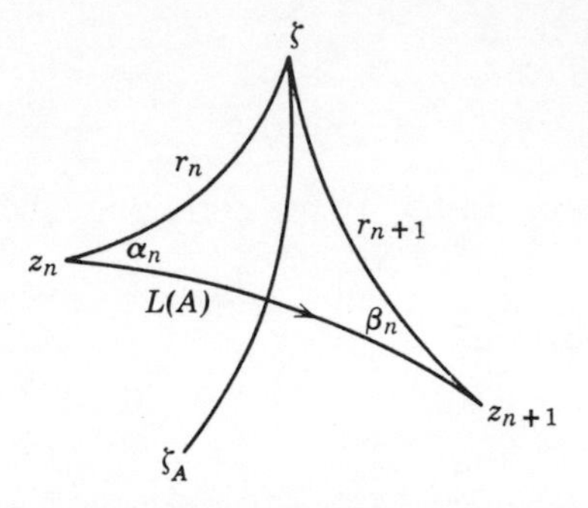

Figure 20

perpendicular bisector M of the side from z_n to z_{n+1} (Figure 21). Since $\delta(\zeta, z_{n+1}) > \delta(\zeta, z_n)$, z_n and ζ lie on the same side of M, and, consequently, M meets the side of the triangle joining ζ to z_{n+1} at an interior point η. Reflection in M carries the noneuclidean triangle with vertices z_n, z_{n+1}, η into itself so that the angle λ at the vertex z_n is equal to the angle β_n at z_{n+1}. On the other hand, we have $\lambda < \alpha_n$, and so $\alpha_n > \beta_n$. Since $\alpha_n + \beta_n + \gamma_n < \pi$, we have, surely, $2\beta_n < \alpha_n + \beta_n < \pi$, and β_n is an acute angle for infinitely many natural n. This implies, for $n \to \infty$, the assertion (6). If there is no minimal m, then we show, analogously, that $r_m > r_{m+1}$ for infinitely many negative m, and this implies $\beta_m > \alpha_m$, $2\alpha_m < \pi$ and

$$\alpha_{-\infty} \leqslant \frac{\pi}{2}. \tag{7}$$

If there is a maximal n, then we denote it by q; otherwise, we put $q = \infty$. Similarly, if there is a minimal m, then we denote it by p; otherwise, we put $p = -\infty$. Then (3) is also valid for $m = p$ and $n = q$ and, by (6) and (7), we have, invariably,

$$\pi - \alpha_q - \beta_q \geqslant 0. \tag{8}$$

Now we consider the finite or countably infinite class of boundary components of $\mathfrak{F}$. In view of the convexity of $\mathfrak{F}$, the triangles $\mathfrak{D}_k$ associated with different boundary components do not overlap, so that the inequalities (4) and (5) with $m = p$, $n = q$ remain in force even if we sum over all the

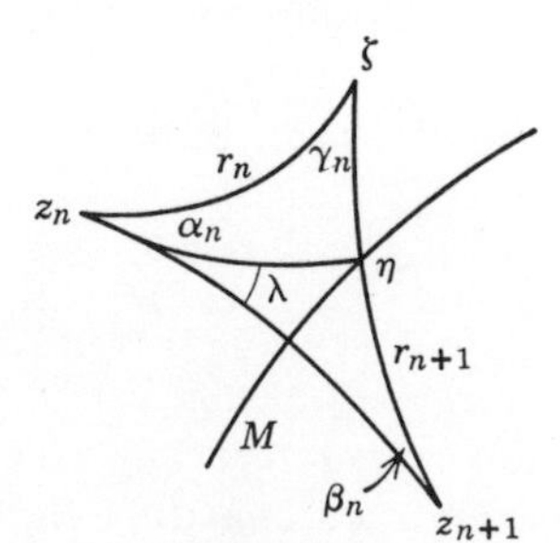

Figure 21

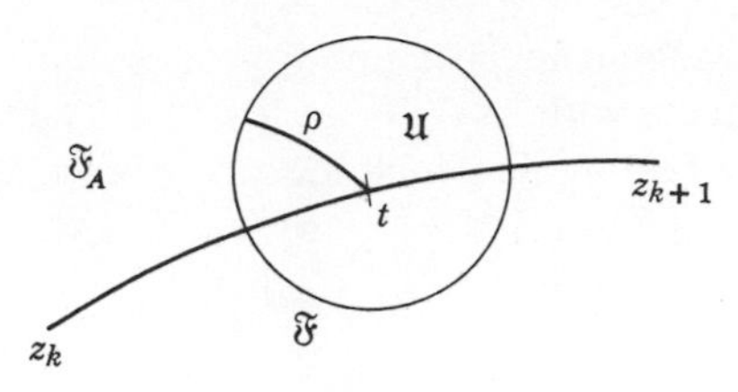

Figure 22

boundary components on the left-hand sides. Now, with (8) in mind, (3) implies the convergence of the sum, over all the proper vertices of $\mathfrak{F}$, of the exterior angles $\pi - \omega_k$. It follows that for every positive ε the number of proper vertices at which the angle is smaller than $\pi - \varepsilon$ is finite. In particular, all of these interior angles have a lower bound θ and only a finite number of them are smaller than $(3\pi)/4$.

IV. We now investigate once more the neighborhood of a boundary point t of $\mathfrak{F}$ in $\mathfrak{E}$ and consider, in this connection, the disk $\mathfrak{U}$ defined by the inequality $\delta(z, t) < \rho$. If $\mathfrak{U}$ contains a point z of the image domain $\mathfrak{F}_A$, then $z_{A^{-1}}$ is in $\mathfrak{F}$ and we have

$$(9) \qquad \delta(\zeta, z_{A^{-1}}) \leqslant \delta(\zeta_{A^{-1}}, z_{A^{-1}}), \quad \delta(\zeta_A, z) \leqslant \delta(\zeta, z),$$

so that, using the noneuclidean triangle inequality,

$$\delta(\zeta, \zeta_A) \leqslant \delta(\zeta, z) + \delta(\zeta_A, z) \leqslant 2\delta(\zeta, z) < 2(\delta(\zeta, t) + \rho).$$

Since the right-hand side of the last inequality is independent of A, the number of possibilities for A is finite. Now, $\mathfrak{F}_A$ is closed in $\mathfrak{E}$ and so ρ can be chosen so small that the intersections of the image regions $\mathfrak{F}_A$ with $\mathfrak{U}$ all have t for a boundary point. Let $\mathfrak{S}(A)$ denote the intersection of $\mathfrak{F}_A$ and $\mathfrak{U}$. If z tends to t while restricted to $\mathfrak{S}(A)$, then, by (9), $\delta(\zeta_A, t) \leqslant \delta(\zeta, t)$, except that now, with t in $\mathfrak{F}$, this last relation must hold with an equality sign. If A is not the identity this means that t lies on the perpendicular bisector $L(A)$. Now, t is not a vertex of $\mathfrak{F}$ if and only if there is a single perpendicular bisector $L(A)$ passing through t, in which case, $\mathfrak{U}$ lies in the union of $\mathfrak{F}$ and $\mathfrak{F}_A$. If, in addition, t is on the boundary component R between the vertices z_k and z_{k+1}, then $\mathfrak{F}$ and $\mathfrak{F}_A$ abut precisely along the noneuclidean segment from z_k to z_{k+1} (Figure 22). Now let t be a proper vertex z_k. For all relevant A, $t_{A^{-1}}$ is a boundary point of $\mathfrak{F}$. If this point were not a vertex of $\mathfrak{F}$, then it would be the point of contact of $\mathfrak{F}$ and of exactly one image region $\mathfrak{F}_B$. But then the image regions $\mathfrak{F}_A$ and $\mathfrak{F}_{AB}$ would form a full neighborhood of t in contradiction to the fact that t is a vertex of $\mathfrak{F}$. It follows that $t_{A^{-1}}$ is a vertex of $\mathfrak{F}$, and there is associated with it an interior angle ω belonging to the interval $\theta \leqslant \omega < \pi$. The conformal nature of the mapping A and the noneuclidean convexity of $\mathfrak{F}$ imply that $\mathfrak{S}(A)$ is a noneuclidean circular sector with angle ω and vertex t. If $\mathfrak{F}_{A_1}, \ldots, \mathfrak{F}_{A_g}$ are all the image regions, including $\mathfrak{F}$, which meet at t, then we have for the corresponding angles $\omega = \omega^{(k)}$ $(k = 1, \ldots, g)$ the formula

$$\omega^{(1)} + \cdots + \omega^{(g)} = 2\pi; \tag{10}$$

this is so because the g sectors $\mathfrak{S}(A_k)$ fill the neighborhood $\mathfrak{U}$ of t (Figure 23). We also obtain the estimate $g \leqslant 2\pi\theta^{-1}$, which, in addition to proving once more the finiteness of g, yields an upper bound for g that is independent of the vertex t.

We now show that $\mathfrak{F}$ has only finitely many proper vertices. First we assume that Γ contains no elliptic elements so that there are no fixed points of Γ. Then the g points $t_{A^{-1}}$ $(A = A_1, \ldots, A_g)$ are equivalent to t and distinct; in fact, for different mappings A and B, the mapping $B^{-1}A$ is not the identity and so

$$t_{A^{-1}} \neq (t_{A^{-1}})_{B^{-1}A} = t_{B^{-1}}.$$

On the other hand, these points are proper vertices of $\mathfrak{F}$ with associated interior angles $\omega^{(k)}$ $(k = 1, \ldots, g)$. Since the g regions $\mathfrak{F}_A$ yield the g sectors with the angles $\omega^{(k)}$ which fill the neighborhood $\mathfrak{U}$ of t, there can be no other boundary points of $\mathfrak{F}$ equivalent to t. It follows that proper vertices of $\mathfrak{F}$ form a finite or countably infinite number of classes of equivalent points, each class containing three or more, but always finitely many, points and the associated angles adding up to 2π. Now, the number of angles smaller than $(3\pi)/4$ is finite and, since it is not possible to obtain the sum 2π by addition of angles belonging to the interval $(3\pi)/4 \leqslant \omega < \pi$, we conclude that the number of classes is finite. Turning to the general case when Γ contains elliptic mappings, we suppose that there are exactly h distinct mappings $C = C_1, \ldots, C_h$ in Γ which carry the vertex t of $\mathfrak{F}$ into itself; here h is invariably $\geqslant 1$, for one of these mappings is the identity. If $h > 1$, then it is

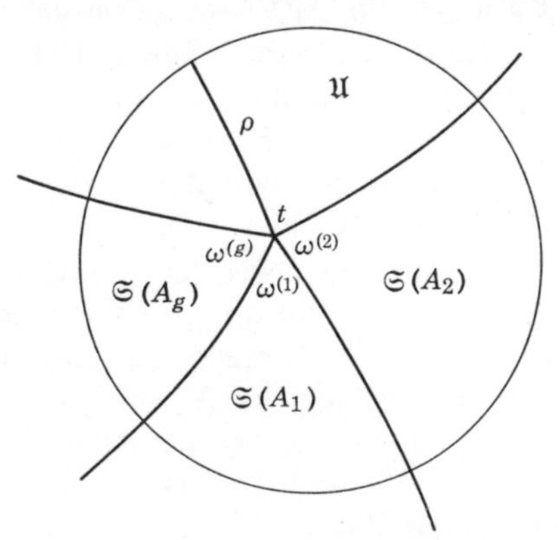

Figure 23

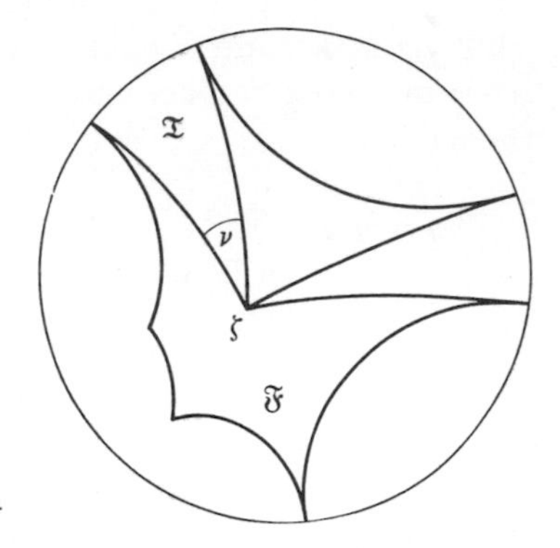

Figure 24

no longer true that the g points $t_{A^{-1}}$ ($A = A_1, \ldots, A_g$) are all distinct; in fact, as we are about to show, they are equal in groups of h. Specifically, for a fixed A, the h different mappings CA ($C = C_1, \ldots, C_h$) carry the point $t_{A^{-1}}$ into t, and the h corresponding image regions $\mathfrak{F}_{CA}$ all have the point t in common, so that h of the points $t_{A_1^{-1}}, \ldots, t_{A_g^{-1}}$ coincide with $t_{A^{-1}}$. Conversely, the equality

$$t_{A_k^{-1}} = t_{A^{-1}}$$

implies that t is a fixed point of the mapping $A_k A^{-1} = C$ and so $A_k = CA$. This shows that the points in question are equal in groups of h, as asserted, and therefore the angles $\omega^{(1)}, \ldots, \omega^{(g)}$ must also be equal in groups of h. If we divide the proper vertices of $\mathfrak{F}$ once again into equivalence classes, then the sum of the angles at the different vertices equivalent to t is $2\pi h^{-1}$, and this value is again not the sum of angles in the interval $(3\pi)/4 \leqslant \omega < \pi$. This implies the finiteness of the number of proper vertices in the general case.

Finally we show that $\mathfrak{F}$ has finitely many boundary components. Since the number of proper vertices is finite, p and q are finite so that $\alpha_p = \beta_q = 0$ and $\pi - \alpha_p - \beta_q = \pi$. When we sum in (3), with $m = p$ and $n = q$, over all the boundary components, then each boundary component contributes at least π to the right-hand side while the left-hand side stays bounded. It follows that the number of R is finite and this means that the number of improper vertices of $\mathfrak{F}$ is also finite.

V. We join the center ζ of $\mathfrak{F}$ to all the improper vertices by means of noneuclidean half lines. Now, either $\mathfrak{F}$ is a noneuclidean polygon or two such half lines bound a full noneuclidean sector $\mathfrak{T}$ of $\mathfrak{E}$ with angle ν (Figure 24). If we carry ζ into 0 by a transformation in Ω and introduce polar coordinates, we obtain for the noneuclidean area of the sector the divergent integral

$$I(\mathfrak{T}) = 4\iint_{\mathfrak{T}} \frac{dx\,dy}{(1 - |z|^2)^2} = 4\int_0^1\int_0^\nu \frac{r\,dr\,d\varphi}{(1 - r^2)^2} = 2\nu\int_0^1 \frac{dt}{(1 - t)^2}\,.$$

This yields a contradiction. It follows that $\mathfrak{F}$ is a noneuclidean polygon some of whose vertices may be improper. This completes the proof of the theorem.

Theorem 2: The noneuclidean area of a normal fundamental region of a circle group is at least $\pi/21$.

Proof: In view of Theorem 1, it suffices to consider the case when the fundamental region $\mathfrak{F}$ is a noneuclidean polygon. Let $z_1, \ldots, z_s$ be a complete system of inequivalent proper vertices of $\mathfrak{F}$. Let there be g_k distinct vertices of $\mathfrak{F}$ equivalent to z_k ($k = 1, \ldots, s$) and let $2\pi h_k^{-1}$ be the sum of the g_k associated polygon angles, where h_k is necessarily a natural number. In the case $g_k = 1$, the two sides of $\mathfrak{F}$ which touch at z_k have the same noneuclidean length. Furthermore, let u be the number of improper vertices of $\mathfrak{F}$. Since the number of vertices of $\mathfrak{F}$ is $u + g_1 + \cdots + g_s$ and the sum of its angles is $2\pi(h_1^{-1} + \cdots + h_s^{-1})$, Theorem 4, Section 2, yields the formula

$$\pi^{-1}I(\mathfrak{F}) = u - 2 + \sum_{k=1}^{s}\left(g_k - \frac{2}{h_k}\right). \tag{11}$$

Since every angle of $\mathfrak{F}$ is smaller than π, we have $g_k\pi > 2\pi h_k^{-1}$, and so $g_k h_k > 2$ and

$$g_k - \frac{2}{h_k} \geqslant \begin{cases} 1/3 \\ 1/2 & (h_k > 3) \\ 1 & (g_k > 1). \end{cases} \tag{12}$$

An additional provision concerning our notation is that $g_1 \geqslant g_2 \geqslant \cdots \geqslant g_s$ and $g_k = g_{k+1}$ implies $h_k \geqslant h_{k+1}$.

If we put

$$u - 2 + \sum_{k=1}^{s}\left(g_k - \frac{2}{h_k}\right) = \mu,$$

then, in view of (11), the rational number μ is positive. In view of (12), $\mu \geqslant 1/6$ for $s > 5$, for $s > 1$, $g_1 > 1$ and for $s > 2$, $u > 0$. In the case $s = 5$, $u = 0$, $g_1 = 1$, we have $\mu \geqslant -2 + 3(1 - 2/4) + 2(1 - 2/3) = 1/6$ for $h_3 > 3$ and $\mu = -2 + (1 - 2/h_1) + (1 - 2/h_2) + 3(1 - 2/3) = 1 - 2/h_1 - 2/h_2 \geqslant 1/21$ for $h_3 = 3$. In the case $s = 4$, $u = 0$, $g_1 = 1$, $\mathfrak{F}$ is a noneuclidean rhombus, since its four sides are of equal length, and so $h_1 = h_2$, $h_3 = h_4$, $\mu = -2 + 2(1 - 2/h_1) + 2(1 - 2/h_3) = 2 - 4(1/h_1 + 1/h_3) \geqslant 2/21$. Now let $s = 3$, $u = 0$. For $g_1 > 3$, or $g_2 > 1$ we have $\mu \geqslant 1/3$; for $g_1 = 2$ or 3 and $g_2 = 1$ we have $\mu = a - 2(1/h_2 + 1/h_3)$ with $a = 1$ or $4/3$ or $\geqslant 3/2$, so that $\mu \geqslant 1/21$; for $g_1 = 1$, $\mathfrak{F}$ is a noneuclidean equilateral triangle so that $\mu = 1 - 6/h_1 \geqslant 1/7$. In the case $s = 2$, we have $\mu = u - 2 + g_1 - 2/h_1 + g_2 - 2/h_2 \geqslant 1/21$. In the case $s = 1$ we have $\mu = u - 2 + g_1 - 2/h_1 \geqslant 1/3$. In the case $s = 0$ we have $\mu = u - 2 \geqslant 1$. It follows that, invariably, $\mu \geqslant 1/21$. Now (11) implies our assertion.

We now consider the question of necessary and sufficient conditions for $\mathfrak{F}$ to be a noneuclidean polygon without improper vertices.

Theorem 3: A normal fundamental region $\mathfrak{F}$ of a circle group Γ is compact in $\mathfrak{E}$ if and only if $I(\mathfrak{F})$ is finite and Γ has no parabolic elements.

Proof: By Theorem 2, $\mathfrak{F}$ is a noneuclidean polygon if and only if $I(\mathfrak{F})$ is finite. We now conduct the proof in two steps.

I. Let $I(\mathfrak{F})$ be finite and $\mathfrak{F}$ not compact in $\mathfrak{E}$. Then the noneuclidean polygon $\mathfrak{F}$ has an improper vertex t at which two sides L_1 and M of $\mathfrak{F}$ meet (Figure 25). If L_1 lies on the perpendicular bisector $L(A_1)$, then $\mathfrak{F}$ abuts along L_1 on the image polygon $\mathfrak{F}_{A_1}$ which also has t as an improper vertex. If L_2 is the other side of $\mathfrak{F}_{A_1}$ which terminates at t, then, in turn, $\mathfrak{F}_{A_1}$ abuts along L_2 on an image polygon $\mathfrak{F}_{A_2}$ with improper vertex t. Since all polygon angles at t are 0, we obtain in this way an infinite sequence of adjoining distinct image polygons $\mathfrak{F}, \mathfrak{F}_{A_1}, \mathfrak{F}_{A_2}, \ldots$ with common vertex t. Then all of the image points $t_{A^{-1}}$ $(k = 1, 2, \ldots)$ are improper vertices of $\mathfrak{F}$ and only finitely many of them are distinct. This means that there exist among the A_k two distinct mappings A and B with $t_{A^{-1}} = t_{B^{-1}}$, $t = t_{AB^{-1}}$, so that t is a fixed point of the mapping $AB^{-1} = C$ which is not the identity and belongs to Γ. Since t lies on $|z| = 1$, C is hyperbolic or parabolic.

If C is hyperbolic, then let u be the other fixed point of C (Figure 26). We join t and u by a noneuclidean line G called the *axis of* C since it is carried by C into itself. We also join t and u by two equidistant curves G_1 and G_2 on either side of G. We recall that G_1 and G_2 are euclidean circular arcs. Finally, we introduce a perpendicular H to G and its image H_C under

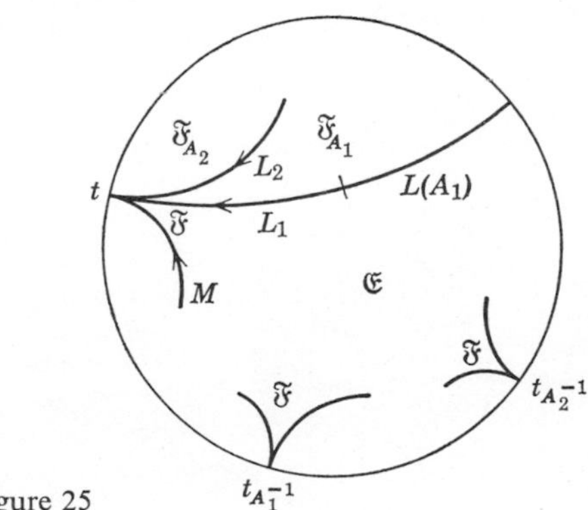

Figure 25

the mapping C. H and H_C determine on the strip $\mathfrak{S}$ formed by G_1 and G_2 a circular quadrilateral $\mathfrak{B}$ which is a fundamental region on $\mathfrak{S}$ of the infinite cyclic subgroup of Γ generated by C. Since the sides L_1 and M of $\mathfrak{F}$ are perpendicular to the unit circle at t, the intersection of $\mathfrak{F}$ and $\mathfrak{S}$ contains a sequence of points ξ converging to t. By its definition, $\mathfrak{B}$ invariably contains a point η equivalent to ξ, and, in view of the compactness of $\mathfrak{B}$ in $\mathfrak{E}$, the noneuclidean distance $\delta(\zeta, \eta)$ is bounded; here ζ again denotes the center of $\mathfrak{F}$. In view of the minimality property of the points of $\mathfrak{F}$, we have, on the other hand, $\delta(\zeta, \xi) \leqslant \delta(\zeta, \eta)$. This yields a contradiction, since $\delta(\zeta, \xi)$ increases beyond all bounds as $\xi \to t$. It follows that C is parabolic. This justifies the conclusion that if Γ has no parabolic elements, then $\mathfrak{F}$ is compact in $\mathfrak{E}$.

II. Let $\mathfrak{F}$ be compact. Then $\mathfrak{F}$ is contained in a disk $|z| < r < 1$, and so $I(\mathfrak{F})$ is finite. It follows that $\mathfrak{F}$ is a noneuclidean polygon all of whose vertices are proper. We assume that Γ has a parabolic element C. After mapping $\mathfrak{E}$ by means of a fractional linear substitution onto the upper half plane so that the fixed point of C goes into ∞, we transform the mapping C into the form $w = z + 2c$ with real $c \neq 0$. The choice $z = -c + iy$ yields $w = c + iy$. If we take in addition to z and w the two ends of the noneuclidean line in $\mathfrak{H}$ determined by z and w, that is, the points $\pm\sqrt{c^2 + y^2}$ on the real axis, then the cross ratio of these four points tends to 1 as $y \to \infty$. It follows that $\delta(z, w) \to 0$. This means that it is possible to find a sequence of points z in $\mathfrak{E}$ with $\delta(z, z_C) \to 0$. Let z_A be a point in $\mathfrak{F}$ equivalent to z; here, A is seen to depend on the position of z in $\mathfrak{E}$. In view of the compactness of $\mathfrak{F}$, we may suppose the sequence of the z chosen so that the points z_A tend to a point ξ of $\mathfrak{F}$. Then

$$\delta(z_A, \xi) \to 0 \tag{13}$$

and so

$$\delta(z_{AC}, \xi_{ACA^{-1}}) \to 0. \tag{14}$$

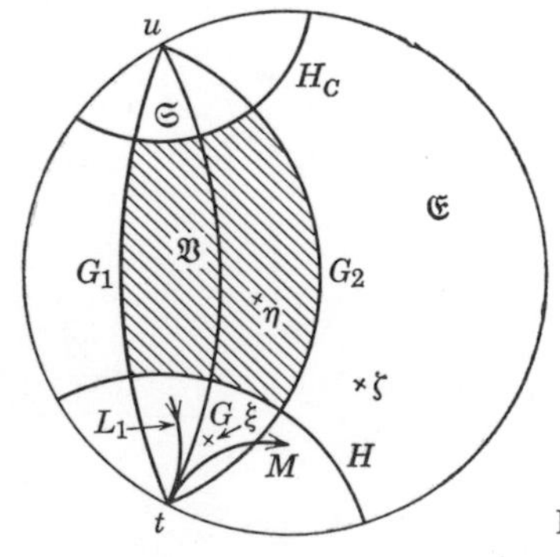

Figure 26

On the other hand,

$$\delta(z_A, z_{AC}) \to 0. \tag{15}$$

After two applications of the triangle inequality, we now obtain, with the aid of (13), (14), and (15), the relation $\delta(\xi, \xi_{ACA^{-1}}) \to 0$, which shows that $\xi_{ACA^{-1}}$ tends to ξ. Since the group Γ is discontinuous in $\mathfrak{E}$, there exists a mapping A with $\xi_{ACA^{-1}} = \xi$. With $\xi_{A^{-1}} = \eta$, η is seen to be a fixed point in $\mathfrak{E}$ of the parabolic mapping C. This contradiction shows that Γ contains no parabolic mapping and so completes the proof.

The results obtained so far suggest the question: for a given circle group, what are the common features of normal fundamental regions with different centers and, more generally, of different fundamental regions? These questions are dealt with in the next two theorems.

Theorem 4: If a circle group Γ has a compact fundamental region then every normal fundamental region of Γ is compact.

Proof: Let $\mathfrak{G}$ be a compact fundamental region of Γ and $\mathfrak{F}$ a normal fundamental region with center ζ. For all points z in $\mathfrak{F}$ and all elements A of Γ, we have the inequality $\delta(\zeta, z) \leqslant \delta(\zeta, z_A)$. Since $\mathfrak{G}$ is compact, the noneuclidean distances between ζ and the points η of $\mathfrak{G}$ are bounded, $\delta(\zeta, \eta) < d$, say. Given z in $\mathfrak{F}$, we select a mapping A in Γ such that $z_A = \eta$ is in $\mathfrak{G}$, and obtain from the two preceding inequalities the required estimate $\delta(\zeta, z) < d$.

Theorems 3 and 4 show that a circle group containing parabolic elements cannot have a compact fundamental region. On the other hand, it is obvious that any circle group has a noncompact fundamental region.

Theorem 5: Any two fundamental regions of a circle group have the same noneuclidean area.

Proof: Let $\mathfrak{F}$ and $\mathfrak{G}$ be two fundamental regions of a circle group Γ. For a positive number $r < 1$, let $\mathfrak{G}_r$ be the closure of the part of $\mathfrak{G}$ lying in the disk $|z| < r$. If A ranges over all the distinct mappings in Γ, then the images $\mathfrak{F}_A$ form a covering of $\mathfrak{G}_r$; we denote the intersection of $\mathfrak{G}_r$ and $\mathfrak{F}_A$ by $\mathfrak{H}(A)$. Since $\mathfrak{H}(A)$ also has euclidean content in the Jordan sense, we can associate with every $\varepsilon(A) > 0$ an open covering of $\mathfrak{H}(A)$ consisting of a finite number of squares $\mathfrak{P}_1(A), \ldots, \mathfrak{P}_m(A)$ and finitely many nonoverlapping squares $\mathfrak{G}_1(A), \ldots, \mathfrak{G}_n(A)$ contained in $\mathfrak{H}(A)$ such that

$$\sum_{k=1}^{m} I(\mathfrak{P}_k(A)) < \varepsilon(A) + \sum_{l=1}^{n} I(\mathfrak{G}_l(A)). \tag{16}$$

In addition to A, these squares and the numbers m and n of the squares depend on r. Both sums in (16) approximate the noneuclidean area $I(\mathfrak{H}(A))$

from below and from above with an error whose absolute value is less than $\varepsilon(A)$. We associate such approximations with each of the mappings A (whose number, we recall, is finite or countably infinite) and make the additional stipulation that the sum of all the $\varepsilon(A)$ is smaller than an arbitrary positive number ε. By the covering theorem it is possible to cover the compact region $\mathfrak{G}_r$ by means of a finite number of squares $\mathfrak{P}$. If we sum over A, then the left-hand side of (16) yields an upper bound S for $I(\mathfrak{G}_r)$ and the right-hand side yields a corresponding lower bound T for $\varepsilon + I(\mathfrak{G}_r)$. Since, on the other hand, we have the inequalities

$$S \geqslant \sum_A I(\mathfrak{H}(A)), \qquad T < \varepsilon + \sum_A I(\mathfrak{H}(A)), \qquad S < T,$$

we obtain, letting $\varepsilon \to 0$, the formula

$$\sum_A I(\mathfrak{H}(A)) = I(\mathfrak{G}_r),$$

which could have been obtained with the aid of the Lebesgue theory without the use of the estimates above. Now, $I(\mathfrak{H}(A))$ is also the noneuclidean area of the intersection of $\mathfrak{F}$ and $(\mathfrak{G}_r)_{A^{-1}}$. Since $\mathfrak{G}_r$ is part of a fundamental region, the individual image regions $(\mathfrak{G}_r)_{A^{-1}}$ do not overlap, and the same holds for the corresponding intersections. It follows that

$$\sum_A I(\mathfrak{H}(A)) \leqslant I(\mathfrak{F}), \qquad I(\mathfrak{G}_r) \leqslant I(\mathfrak{F}).$$

Now let $r \to 1$. Then we obtain the inequality $I(\mathfrak{G}) \leqslant I(\mathfrak{F})$. Interchanging $\mathfrak{F}$ and $\mathfrak{G}$ yields the assertion of the theorem.

Theorem 5 shows that the noneuclidean area $I(\mathfrak{F})$ depends only on the circle group Γ and not on the choice of fundamental region. This justifies the notation $I(\Gamma)$, which we shall use in the sequel. If $I(\Gamma)$ is finite, then, in view of Theorem 1, we can choose as fundamental region a noneuclidean polygon $\mathfrak{F}$. In that case we shall call Γ a *polygon group* and $\mathfrak{F}$ a *normal polygon*. If, in addition, Γ contains no parabolic elements, then all the vertices of $\mathfrak{F}$ are proper. This is true, in particular, for hyperbolic polygon groups. We are about to give a special geometric interpretation of the noneuclidean area of a normal polygon $\mathfrak{F}$ in this particularly important case. On each side L of $\mathfrak{F}$ there abuts exactly one image region $\mathfrak{F}_A$, so that $L_{A^{-1}}$ is also a side of $\mathfrak{F}$ with abutting image region $\mathfrak{F}_{A^{-1}}$. If $L_{A^{-1}}$ and L coincided, then A and A^{-1} would be the same and A^2 would be the identity, contradicting the assumed hyperbolic character of Γ. Since in this argument we can interchange the sides L and $L_{A^{-1}}$, the sides of $\mathfrak{F}$ can be grouped in equivalent pairs, n in number, so that $\mathfrak{F}$ has $2n$ sides. Transition from $\mathfrak{F}$ to $\mathfrak{F}_{A^{-1}}$ preserves orientation, so that L and $L_{A^{-1}}$ in $\mathfrak{F}$ have opposite orientations (Figure 27). Furthermore, since no vertex is a fixed point,

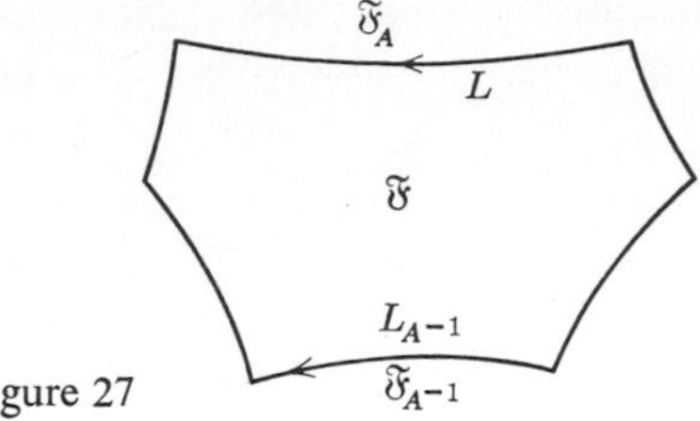

Figure 27

L and $L_{A^{-1}}$ have no common end point. If we identify equivalent boundary points, that is, if we join equivalent sides in the proper sense, then $\mathfrak{F}$ goes over into a closed surface $\mathfrak{F}^*$. As was proved in IV of Theorem 1, the angles at the vertices of $\mathfrak{F}$ which form a single point of $\mathfrak{F}^*$ add up, by (10), to 2π. If we regard $\mathfrak{F}^*$ as a polyhedron, we can apply the formula of Euler and Poincaré mentioned in Section 4 of Chapter 2 (Volume 1) which asserts that the connection between the numbers E, K, F of vertices, edges and faces is

$$E - K + F = 2 - 2p, \tag{17}$$

where p is the genus of the polyhedron. If ω_k ($k = 1, \ldots, 2n$) varies over the angles of $\mathfrak{F}$, then summation over the vertices of $\mathfrak{F}$ yields the formula

$$2\pi E = \sum_{k=1}^{2n} \omega_k,$$

where $K = n$ and $F = 1$. By (17),

$$4(p-1)\pi = (2n-2)\pi - \sum_{k=1}^{2n} \omega_k,$$

and, since $2n$ is the number of sides of the polygon, Theorem 4 of Section 2 yields for the noneuclidean area of $\mathfrak{F}$ the value

$$I(\Gamma) = I(\mathfrak{F}) = 4(p-1)\pi. \tag{18}$$

In view of Theorem 5, this shows that the genus of the closed surface $\mathfrak{F}^*$ obtained by joining equivalent sides of $\mathfrak{F}$ is independent of the center of the normal polygon $\mathfrak{F}$—a result which could also be proved by making use of the topological invariance of the genus.

Just as the Riemann surface of an algebraic function, so too the surface $\mathfrak{F}^*$ is compact, and just as the polygon $\mathfrak{N}_p$ obtained in Section 2 of Chapter 2 (Volume I) by dissection of such a Riemann surface, so too the sides of $\mathfrak{F}$ are associated in pairs. It must be remembered, however, that $\mathfrak{F}^*$ does not lie over the complex number sphere but is obtained from the surface $\mathfrak{F}$ in the z-plane by identification of equivalent boundary points. By bringing in the connection between $\mathfrak{E}$, Γ, and $\mathfrak{F}$, we could easily generalize our earlier definition of a Riemann surface so as to include $\mathfrak{F}$, but this is of no importance for our future objectives. The surface $\mathfrak{F}^*$ will play a role in the sequel.

5. Poincaré series

The main purpose of this chapter is the investigation of functions of z meromorphic in $\mathfrak{E}$ which are invariant under the substitutions of a hyperbolic polygon group. By way of preparation for this task we introduce certain regular functions whose quotients will be the required functions. Let Γ be a circle group of substitutions

$$(1) \qquad w = z_A = \frac{\bar{a}z + \bar{b}}{bz + a}, \qquad A = \begin{pmatrix} \bar{a} & \bar{b} \\ b & a \end{pmatrix},$$

with $a\bar{a} - b\bar{b} = 1$. We call a regular function $f(z)$ in $\mathfrak{E}$ an *automorphic form of weight l* associated with the group Γ if, for all A in Γ, $f(z)$ satisfies the transformation rule

$$(2) \qquad f(z_A) = (bz + a)^l f(z)$$

identically in z. Since the substitution $w = z_A$ is not affected by the replacement of the matrix A by $-A$, l may be supposed an even integer. By (14) in Section 1,

$$(3) \qquad 1 - |z_A|^2 = |bz + a|^{-2} (1 - |z|^2),$$

so that

$$(1 - |z|^2)^{l/2} \, |f(z)| = g(z)$$

is a real nonnegative function of z invariant under Γ. In particular, if Γ has a compact fundamental region $\mathfrak{F}$, then $g(z)$ has an absolute maximum μ at a point ξ of $\mathfrak{F}$, and, in view of the invariance of $g(z)$ under Γ, μ is also the maximum on $\mathfrak{E}$. It follows that

$$|f(z)| \leqslant \mu(1 - |z|^2)^{-l/2}$$

on $\mathfrak{E}$. If l is negative, then, for $z \rightarrow 1$, $f(z)$ converges uniformly to 0 and, by the maximum principle, $f(z) = 0$. If $l = 0$, then $f(z)$ takes on a maximum at ξ, so that the function $f(z)$ is constant. Hence, for a compact fundamental region, the positive values of l are the only values of interest.

Our problem is to construct nonconstant automorphic forms. This is done by means of certain series introduced by Poincare and named after him.

Theorem 1: Let Γ be a circle group. The sum

$$h(z, s) = \sum_A |bz + a|^{-s} \qquad (|z| < 1),$$

taken over all the mappings A in Γ, converges for real $s > 2$. If δ is positive and $r < 1$, then this convergence is uniform in z and s for $|z| \leqslant r$ and $s \geqslant 2 + \delta$.

Proof: We use an important idea which goes back to Poincaré. We start with a noneuclidean circle in a normal fundamental region $\mathfrak{F}$ with center at the center ζ of $\mathfrak{F}$. The circle bounds the region $\mathfrak{G}$. The image regions $\mathfrak{G}_A$ of $\mathfrak{G}$ lie in the unit circle and are disjoint, so that for the sum of their euclidean areas we have

$$\sum_{A} \iint_{G_A} dx\, dy \leqslant \pi. \tag{4}$$

If we put $w = u + iv$, $z = x + iy$, then the functional determinant of the mapping (1) is given by

$$\frac{d(u, v)}{d(x, y)} = \frac{d(w, \bar{w})}{d(z, \bar{z})} = |bz + a|^{-4}, \tag{5}$$

so that

$$\iint_{\mathfrak{G}_A} du\, dv = \iint_{\mathfrak{G}} |bz + a|^{-4}\, dx\, dy. \tag{6}$$

Since $|a|^2 - |b|^2 = 1$, we have $|b| < |a|$ and, for $|z| \leqslant r < 1$, $|\xi| \leqslant r$, we have

$$\frac{|bz + a|}{|b\xi + a|} \leqslant \frac{|a| + |a|\, r}{|a| - |a|\, r} = \frac{1 + r}{1 - r} \leqslant (1 - r)^{-2}. \tag{7}$$

Now we choose r so close to 1 that $\mathfrak{G}$ lies entirely in $|z| \leqslant r$ and denote the euclidean area of $\mathfrak{G}$ by γ. In view of (4), (6), and (7), with $\xi = 0$, we obtain the estimate

$$\gamma(1 - r)^8 \sum_{A} |a|^{-4} \leqslant \sum_{A} \iint_{\mathfrak{G}} |bz + a|^{-4}\, dx\, dy \leqslant \pi.$$

It follows that the series $h(0, 4)$ converges. Interchanging z and ξ in (7) and putting $\xi = 0$, we obtain

$$|bz + a|^{-s} \leqslant (1 - r)^{-2s}\, |a|^{-s} \qquad (s > 0). \tag{8}$$

Since the Dirichlet series $(1 - r)^{-2s} h(0, s)$ converges for $s = 4$, it converges uniformly for $s \geqslant 4$. In view of (8), it follows that the series $h(z, s)$ converges uniformly in z and s for $|z| \leqslant r$, $s \geqslant 4$.

To prove the assertion completely, we modify the idea as follows. Introducing polar coordinates we see that the integral

$$p(\alpha) = \iint_{\mathfrak{T}} (1 - |z|^2)^{\alpha}\, dx\, dy = \int_0^1 \int_0^{2\pi} (1 - r^2)^{\alpha} r\, dr\, d\varphi = \pi \int_0^1 (1 - x)^{\alpha}\, dx$$

converges for $\alpha > -1$ and diverges for $\alpha \leqslant -1$. By transforming the corresponding integral over $\mathfrak{G}_A$ to an integral over $\mathfrak{G}$, we obtain, using (3) and (5), the inequality

$$\sum_A \iint_{\mathfrak{G}} (1 - |z|^2)^\alpha \, |bz + a|^{-2\alpha-4} \, dx \, dy \leqslant p(\alpha) \qquad (\alpha > -1).$$

In $\mathfrak{G}$ we have $1 - r^2 \leqslant 1 - |z|^2$. Using the earlier reasoning, we deduce the convergence of the series $h(0, s)$ for $s = 2\alpha + 4 > 2$, and from this the uniform convergence of $h(z, s)$ for $|z| \leqslant r$, $s \geqslant 2 + \delta$. This completes the proof.

If Γ has a compact fundamental region $\mathfrak{F}$, then we can use the same idea to prove that the series $h(z, s)$ diverges for every z in $\mathfrak{E}$ and every real $s \leqslant 2$. To this end we choose $r < 1$ such that $\mathfrak{F}$ lies in the circle $|z| \leqslant r$ and suppose $\alpha = s/2 - 2 \leqslant -1$. Then

$$p(\alpha) = \sum_A \iint_{\mathfrak{F}_A} (1 - |z|^2)^\alpha \, dx \, dy = \sum_A \iint_{\mathfrak{F}} (1 - |z|^2)^\alpha \, |bz + a|^{-s} \, dx \, dy$$

diverges, and for the integrand on the right we have by (7) the estimate

$$(1 - |z|^2)^\alpha \, |bz + a|^{-s} \leqslant (1 - r^2)^\alpha (1 - r)^{-2|s|} \, |a|^{-s}.$$

From this we deduce the divergence of the sum $h(0, s)$ for $s \leqslant 2$ and, more generally, by (7), the divergence of $h(z, s)$.

By Theorem 1 the sum

$$f_k(z) = \sum_A (bz + a)^{-2k} \qquad (k = 2, 3, \ldots) \tag{9}$$

converges uniformly for every k and $r < 1$ in the disk $|z| \leqslant r$. Since $bz + a$ is different from 0 in $\mathfrak{E}$, the general term of the sum in (9) is regular in the unit disk. By Weierstrass' theorem, the same is true of the function $f_k(z)$. To investigate the behavior of $f_k(z)$ under the substitutions of Γ, we make use of the relation

$$\frac{dw}{dz} = \frac{dz_A}{dz} = (bz + a)^{-2},$$

which enables us to set

$$f_k(z) = \sum_A \left(\frac{dz_A}{dz}\right)^k.$$

If B is a fixed element of Γ, then we have

$$\frac{dz_{AB}}{dz} = \frac{dz_{AB}}{dz_B} \frac{dz_B}{dz}.$$

As A varies over Γ so does AB. In view of the absolute convergence of the sums, this implies

$$\left(\frac{dz_B}{dz}\right)^k f_k(z_B) = \left(\frac{dz_B}{dz}\right)^k \sum_A \left(\frac{dz_{AB}}{dz_B}\right)^k = \sum_A \left(\frac{dz_{AB}}{dz}\right)^k = \sum_A \left(\frac{dz_A}{dz}\right)^k = f_k(z),$$

and, therefore, also

$$f_k(z_A) = (bz + a)^{2k} f_k(z) \qquad (k = 2, 3, \ldots). \tag{10}$$

In accordance with our definition, (10) shows $f_k(z)$ to be an automorphic form of weight $2k$ associated with the circle group Γ.

The series (9) are called *Poincaré series of special type*. *Poincaré series* are series of the form

$$\varphi_k(z) = \sum_A (bz + a)^{-2k} \varphi(z_A) \qquad (k = 2, 3, \ldots),$$

where $\varphi(z)$ is a regular bounded function on $\mathfrak{E}$. Since $\varphi(z_A)$ is bounded uniformly in z and A for $|z| < 1$, we deduce, as above, the uniform convergence of the series for $|z| \leqslant r < 1$ and the regularity of the function $\varphi_k(z)$ for $|z| < 1$. The proof of (10) also carries over to this series. Thus for every admissible choice of $\varphi(z)$ this series represents an automorphic form of weight $2k$ associated with Γ. The Poincaré series of special type will be adequate for our purposes. With Γ supposed hyperbolic, we now prove two simple theorems about the associated Poincaré series $f_k(z)$.

Theorem 2: If Γ is a hyperbolic polygon group, then, for k sufficiently large, none of the associated Poincaré series $f_k(z)$ is identically 0.

Proof: For the identity substitution of Γ we have $a = \pm 1$, $b = 0$. For the remaining substitutions of Γ, $b \neq 0$; otherwise $z = 0$ would be an elliptic fixed point. It follows that $|a| > 1$ for the substitutions of Γ other than the identity. By Theorem 1 and definition (9), the series $f_k(0)$ ($k = 2, 3, \ldots$) converges uniformly in the parameter k. Transition to the limit $k \to \infty$ yields $f_k(0) \to 1$, which implies our assertion.

Theorem 3: If Γ is a hyperbolic polygon group and the Poincaré series $f_k(z)$ is not identically zero, then the rim of the unit disk is the natural boundary of $f_k(z)$.

Proof: Let $\mathfrak{F}$ be a compact fundamental region of Γ, ξ a point of $\mathfrak{F}$, u a point on $|z| = 1$ and L a noneuclidean line with end u. As z varies on L and tends toward u, we traverse infinitely many of the image regions $\mathfrak{F}_{A_l}$ ($l = 1, 2, \ldots$); this follows from the compactness of all the $\mathfrak{F}_A$. Let ρ be the maximum of the noneuclidean distances from ξ to a variable point on $\mathfrak{F}$. The locus of all points whose noneuclidean distance from L is ρ

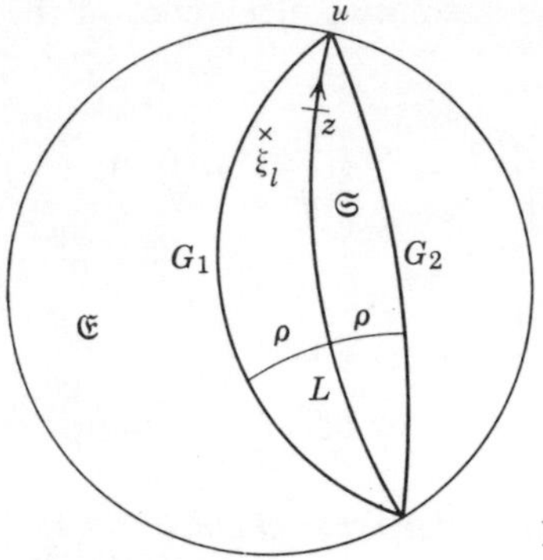

Figure 28

consists of two equidistant curves G_1 and G_2 which bound a strip $\mathfrak{S}$ containing L (Figure 28). Since the noneuclidean distance $\delta(z, \xi_A)$ of every point z of of $\mathfrak{F}_A$ on L is at most ρ, it follows that the images $\xi_{A_l} = \xi_l$ $(l = 1, 2, \ldots)$ of ξ lie in $\mathfrak{S}$. The boundedness of these distances shows that as z tends to u the sequence of points ξ_l can have no limit point other than a point on the rim of the unit disk. Since the rim of the unit disk has no points in common with $\mathfrak{S}$ other than the ends of L, it follows that the points ξ_l also tend to u.

Now let ξ be a point other than a zero of $f_k(z)$ and u a point on $|z| = 1$. If a_l and b_l are the values of a and b corresponding to A_l, then, by (10), we have the relation

$$f_k(\xi_l) = (b_l\xi + a_l)^{2k} f_k(\xi) \qquad (l = 1, 2, \ldots), \tag{11}$$

where, as a result of the convergence of the Poincaré series, the factor $(b_l\xi + a_l)^{2k}$ must become infinite for $l \to \infty$. It follows that for every point u on the unit circle there is a sequence of interior points with limit u for which the functional values of $f_k(z)$ tend to ∞. This completes the proof.

Theorems 2 and 3 imply the existence of nonconstant Poincaré series of special type.

6. *The field of automorphic functions*

Let Γ be a circle group. We call a function $f(z)$ an *automorphic function associated with* Γ if it is meromorphic on $\mathfrak{E}$ and invariant under all the substitutions of Γ. Thus the condition (2) of the preceding section is replaced by the more stringent requirement

$$f(z_A) = f(z) \tag{1}$$

for all A in Γ. Now, however, we assume only that $f(z)$ is meromorphic in $\mathfrak{E}$ and not, as in the case of automorphic forms, regular in $\mathfrak{E}$. It is clear that the quotient of two automorphic forms f_1, f_2 of equal weight is an automorphic function provided that the denominator f_2 does not vanish identically. We shall soon be able to rule out the possibility that these quotients are invariably constant. Again, it is clear that the automorphic functions associated with Γ form a field K. We restrict ourselves to hyperbolic polygon groups. Our aim is to prove that K is an algebraic function field of transcendence degree 1. We shall also prove that it is possible to express every function in K as a rational combination with constant coefficients of a finite number of fixed Poincaré series of special type.

At the end of Section 4 we identified equivalent boundary points of a normal polygon $\mathfrak{F}$ and obtained the closed surface $\mathfrak{F}^*$ by joining equivalent sides. This establishes a one-to-one correspondence between the equivalence classes on $\mathfrak{E}$ and the points of $\mathfrak{F}^*$. Without forming $\mathfrak{F}^*$, it is possible to set up such a correspondence on $\mathfrak{F}$ in the following manner. In each of the n pairs of equivalent sides of $\mathfrak{F}$ we omit, by some definite rule, one side. Of the remaining vertices we retain, again by definite choice, one from each equivalence class. This yields a new region $\mathfrak{F}_0$ which contains exactly one point from each equivalence class on $\mathfrak{E}$; this is true of $\mathfrak{F}$ only for interior points. We note that $\mathfrak{F}_0$ is neither closed nor open in $\mathfrak{E}$.

Theorem 1: Let $f(z)$ be a nonconstant function in K. Consider $f(z)$ on $\mathfrak{F}_0$. If we count multiplicities, then $f(z)$ takes on every value equally often.

Proof: In view of the compactness of $\mathfrak{F}$, the nonconstant meromorphic function $f(z)$ has only finitely many zeros and poles in $\mathfrak{F}_0$. Let p be the number of poles and n the number of zeros. If there are neither poles nor zeros on the positively oriented boundary C of $\mathfrak{F}$, then

$$n - p = \frac{1}{2\pi i} \int_C \frac{f'(z)}{f(z)}\, dz. \tag{2}$$

By (1), the function $f(z)$ and the differential $df(z) = f'(z)\, dz$ are invariant under the mappings A in Γ. On the other hand, C can be divided into pairs of equivalent oppositely oriented sides L and $L_{A^{-1}}$ which correspond to one another under A. It follows that the contributions of the sides of such a pair to the integral in (2) cancel each other, so that $n = p$.

We now generalize the argument above so that zeros and poles can appear on the boundary of $\mathfrak{F}$. If a point ξ of the side L of $\mathfrak{F}_0$ which is not a vertex of $\mathfrak{F}$ is a zero or a pole of the function, then the point $\xi_{A^{-1}}$ of the side $L_{A^{-1}}$ is a zero or a pole of the same multiplicity. In that case we alter the path of integration C in the vicinity of ξ and $\xi_{A^{-1}}$ by going around ξ on a small noneuclidean circular arc B in $\mathfrak{F}_A$, and by going around $\xi_{A^{-1}}$ on the oppositely oriented image arc $B_{A^{-1}}$ in $\mathfrak{F}$ (Figure 29). Similarly, if ξ is a point of $\mathfrak{F}_0$

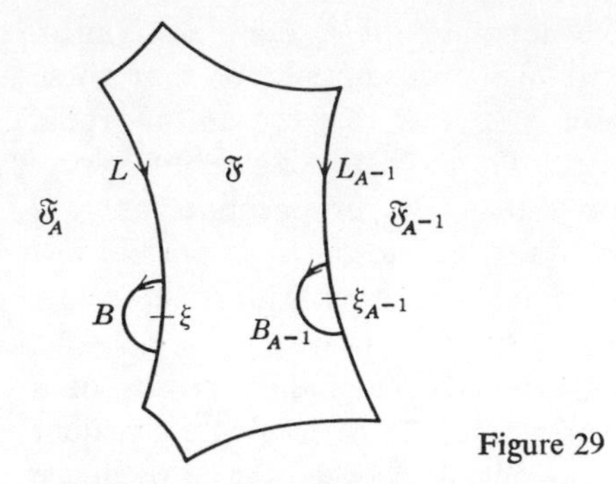

Figure 29

which is a vertex of $\mathfrak{F}$, then we go around ξ on a noneuclidean circular arc B which crosses the image regions $\mathfrak{F}_{A_k}$ $(k = 2, \ldots, g)$ distinct from $\mathfrak{F}$ and meeting at ξ on subarcs B_k, and go around the $g - 1$ other vertices of $\mathfrak{F}$ equivalent to ξ on the oppositely oriented image arcs $(B_k)_{A_k^{-1}}$ in $\mathfrak{F}$ (Figure 30). Applying the formula (2) to the modified boundary curve, we again obtain $n = p$.

Now let c be any constant. Then $f(z) - c = g(z)$ is a nonconstant function in K which has the same poles as $f(z)$. Thus, by what has already been proved, $g(z)$ has exactly p zeros on $\mathfrak{F}_0$. But then $f(z)$ takes on the arbitrarily prescribed value c exactly p times on $\mathfrak{F}_0$. This completes the proof.

Theorem 1 can be formulated without the use of a normal polygon as follows. The number of equivalence classes on $\mathfrak{E}$ in which $f(z)$ takes on the finite or infinite value c is independent of c. We call this number the *order of f* and denote it by $\mathfrak{O}(f)$. If f is constant then we define $\mathfrak{O}(f) = 0$.

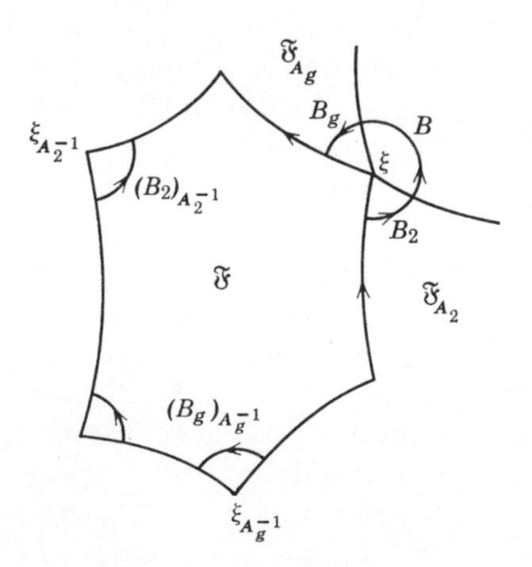

Figure 30

We retain the assumption that $f(z)$ is a nonconstant function in K, that is, that

$$\mathfrak{O}(f) = m > 0.$$

We consider the set of rational functions of f with arbitrary constant coefficients which, clearly, forms a subfield Λ of K. We shall now show that K is an algebraic extension of Λ. As usual, n functions $g_1, \ldots, g_n$ in K are said to be linearly dependent over Λ if the linear homogeneous equation

$$\varphi_1 g_1 + \cdots + \varphi_n g_n \tag{3}$$

has, identically in z, a nontrivial solution $\varphi_1, \ldots, \varphi_n$ consisting of n functions in Λ. In the absence of such a solution the functions $g_1, \ldots, g_n$ are said to be linearly independent over Λ.

Theorem 2: Every $m + 1$ functions in K are linearly dependent over Λ.

Proof: Let $g_0, \ldots, g_m$ be $m + 1$ given functions in K. We put

$$\sum_{k=0}^{m} \mathfrak{O}(g_k) = n \tag{4}$$

and form $m + 1$ polynomials in f with undetermined constant coefficients

$$\varphi_k = c_{k0} f^n + c_{k1} f^{n-1} + \cdots + c_{kn} \qquad (k = 0, \ldots, m). \tag{5}$$

Clearly, the sum

$$\varphi_0 g_0 + \cdots + \varphi_m g_m = h(z)$$

is a function in K which can have poles only at the poles of the $m + 2$ functions g_k $(k = 0, \ldots, m)$ and f. By (4), the g_k have, in all, n poles (counting multiplicities). By (5), the degree of φ_k in f is at most n, so that the order of a pole of φ_k is at most n times the order of the same pole of f. This implies for the order of $h(z)$ the estimate

$$\mathfrak{O}(h) \leqslant n + mn = (m + 1)n,$$

and this estimate is independent of the choice of the coefficients c_{kl} $(k = 0, \ldots, m;\ l = 0, \ldots, n)$. The number of these coefficients is $t = (m + 1)(n + 1)$. We put

$$s = t - 1 = (m + 1)(n + 1) - 1 = (m + 1)n + m > (m + 1)n \tag{6}$$

and choose s points $z_1, \ldots, z_s$ from different equivalence classes at which the functions g_k and f are regular. The requirements

$$h(z_p) = 0 \qquad (p = 1, \ldots, s)$$

determine s homogeneous linear equations for the t unknown coefficients c_{kl}, and, by (6), these equations have a nontrivial solution. This means that

$h(z)$ is a function in K with at least s zeros and at most $(m + 1)n$ poles on $\mathfrak{F}_0$. By (6) and Theorem 1, this is possible only if $h(z)$ vanishes identically in z. Since f is not constant and not all the c_{kl} are zero, not all the φ_k are identically zero. This completes the proof.

By Theorem 2, there is a maximal number r of functions in K which are linearly independent over Λ. The number r is called the *degree of* K *over* Λ. Since the constant 1 is in K, we have, by Theorem 2, the inequality $1 \leqslant r \leqslant m$. We shall see later that $r = m$. If $g_1, \ldots, g_r$ are r functions in K which are linearly independent over Λ, then the linear form

$$\psi_1 g_1 + \cdots + \psi_r g_r = h$$

with arbitrary coefficients $\psi_1, \ldots, \psi_r$ in Λ represents every function h in K exactly once. By a familiar result from algebra, it is possible to choose r constants $c_1, \ldots, c_r$ such that the adjunction of the single function

$$c_1 g_1 + \cdots + c_r g_r = g$$

to Λ yields all of the field K, so that $K = \Lambda(g)$ is a simple algebraic extension of degree r over Λ. We shall return to this point when expressing the generators f and g rationally in terms of Poincaré series of special type. What follows can be presented more briefly using arbitrary Poincaré functions; it is of fundamental importance, however, that the field K can be built up using series of special type only.

The Poincaré series in question,

$$f_l = f_l(z) = \sum_A (bz + a)^{-2l} \qquad (l = 2, 3, \ldots), \tag{7}$$

yield for every value of l an automorphic form of weight $2l$ associated with Γ, which, by Theorem 2 of the preceding section, does not vanish identically for sufficiently large l. We choose $l = q$ so that $f_q(z)$ is not identically zero and put

$$h_k(z) = f_q^{-k} f_{kq} \qquad (k = 1, 2, \ldots). \tag{8}$$

Since f_{kq} and f_q^k are automorphic forms of equal weight $2qk$, it follows that the quotients $h_k(z)$ are automorphic functions associated with Γ. If two points ξ and η satisfy the relation

$$h_k(\xi) \neq h_k(\eta)$$

for at least one index k, then we shall say that *the functions* $h_k(z)$ $(k = 1, 2, \ldots)$ *separate these points.*

We now impose on the given group Γ a further restriction which will later turn out to be inessential. If the matrices

$$A = \begin{pmatrix} \bar{a} & \bar{b} \\ b & a \end{pmatrix} \quad \text{and} \quad A^{-1} = \begin{pmatrix} a & -\bar{b} \\ -b & \bar{a} \end{pmatrix} \qquad (a\bar{a} - b\bar{b} = 1)$$

are mutually inverse elements of Ω, then b and $-b$ have the same absolute value. We call Γ *normalized* if for all A in Γ the pair of mappings A, A^{-1} is uniquely determined by prescribing the absolute value of b alone. In the following theorem we suppose Γ normalized.

Theorem 3: If a point of $\mathfrak{E}$ does not belong to certain finitely many equivalence classes, then the functions $h_k(z)$ separate it from every inequivalent point of $\mathfrak{E}$.

Proof: The proof is in six steps.

I. Let

$$s_1 = \alpha_1 + \alpha_2 + \cdots, \qquad t_1 = \beta_1 + \beta_2 + \cdots \tag{9}$$

be two given absolutely convergent series all of whose terms are different from 0. With $k = 2, 3, \ldots$ we construct the likewise absolutely convergent series

$$s_k = \alpha_1^k + \alpha_2^k + \cdots, \qquad t_k = \beta_1^k + \beta_2^k + \cdots$$

and show that, with proper numbering of the indices, the assumptions

$$s_k = t_k \qquad (k = 1, 2, \ldots) \tag{10}$$

imply the equalities

$$\alpha_l = \beta_l \qquad (l = 1, 2, \ldots). \tag{11}$$

For finite sums in place of (9), the corresponding assertion is a well-known result of algebra. To prove this result for infinite sums, we must make use of the methods of analysis. If τ is a complex variable, then, in view of the absolute convergence of s_1, the series of partial fractions

$$\phi(\tau) = \sum_{l=1}^{\infty} \frac{\alpha_l}{1 - \alpha_l \tau}$$

defines a meromorphic function of τ. All of its poles, namely, the points $\tau = \alpha_l^{-1}$, are simple. If exactly h of the summands in s_1 have the value α_l, then the residue of $\phi(\tau)$ at the corresponding pole is $-h$. If r is the minimum of the absolute values of all the poles, then for $|\tau| < r$ we can expand $\phi(\tau)$ in the following absolutely convergent series in powers of τ:

$$\phi(\tau) = \sum_{l=1}^{\infty} \alpha_l \sum_{k=0}^{\infty} (\alpha_l \tau)^k = \sum_{k=0}^{\infty} s_{k+1} \tau^k.$$

If we put, similarly,

$$\psi(\tau) = \sum_{l=1}^{\infty} \frac{\beta_l}{1 - \beta_l \tau},$$

then (10) implies that the functions $\phi(\tau)$ and $\psi(\tau)$ have the same power series representations and so must coincide. It follows that the functions have the same poles and the same residues at all poles. With the β_l suitably indexed, this implies the assertion (11). At the same time our proof shows that not all the sums s_k $(k = 1, 2, \ldots)$ vanish.

II. Let ξ and η be two inequivalent points of the normal polygon $\mathfrak{F}$ with center $\zeta = 0$. Since the group Γ is hyperbolic, it has no fixed point. It follows that every point of $\mathfrak{E}$ is admissible as the center of a normal fundamental region. Since $f_q(z)$ has only finitely many zeros on $\mathfrak{F}$, the condition

$$f_q(\xi) \neq 0 \tag{12}$$

rules out only finitely many equivalence classes for ξ. Now let

$$h_k(\xi) = h_k(\eta) \qquad (k = 1, 2, \ldots). \tag{13}$$

We suppose the mappings $A = A_l$ $(l = 1, 2, \ldots)$ in Γ indexed, define accordingly $a = a_l$, $b = b_l$, and put

$$f_q(\xi)(b_l\xi + a_l)^{2q} = \alpha_l^{-1} \qquad (l = 1, 2, \ldots),$$

so that, by (7) and (8), we have the equation

$$h_k(\xi) = \sum_{l=1}^{\infty} \alpha_l^k = s_k \qquad (k = 1, 2, \ldots).$$

If we had $f_q(\eta) = 0$, then (8), (12), and (13) would imply $f_{kq}(\eta) = 0$ $(k = 1, 2, \ldots)$ and, by (7), we would obtain a contradiction of the remark at the end of I. Hence $f_q(\eta) \neq 0$, and we can define

$$f_q(\eta)(b_l\eta + a_l)^{2q} = \beta_l^{-1} \qquad (l = 1, 2, \ldots)$$

with

$$h_k(\eta) = \sum_{l=1}^{\infty} \beta_l^k = t_k \qquad (k = 1, 2, \ldots).$$

By (13) and I it follows that, with multiplicities taken into consideration, the infinite sequence of numbers $f_q(\xi)(b_l\xi + a_l)^{2q}$ $(l = 1, 2, \ldots)$ is a permutation of the sequence $f_q(\eta)(b_l\eta + a_l)^{2q}$.

III. We show that $\mathfrak{F}$ consists exactly of the points of $\mathfrak{E}$ satisfying the inequalities

$$|b_l z + a_l| \geqslant 1 \qquad (l = 1, 2, \ldots). \tag{14}$$

By (16) in Section 3, the normal polygon $\mathfrak{F}$ with center $\zeta = 0$ is defined by the conditions

$$\delta(0, z) \leqslant \delta(0, z_A) \qquad (A \text{ in } \Gamma). \tag{15}$$

By (6) in Section 2, the noneuclidean distance $\delta(0, z)$ of a point z from 0 is a monotonically increasing function of its euclidean distance $|z|$. This means that (15) is equivalent to $|z| \leqslant |z_A|$ and

$$1 - |z|^2 \geqslant 1 - |z_A|^2.$$

Our assertion follows by (14) in Section 1.

IV. We suppose the A_l $(l = 1, 2, \ldots)$ indexed so that the absolute values $|b_l|$ form a monotonically increasing sequence. Since Γ is supposed normalized, we can put

$$A_1 = E, \qquad A_{2k+1} = A_{2k}^{-1} \qquad (k = 1, 2, \ldots)$$

with $b_1 = 0$, $b_{2k+1} = -b_{2k}$, $0 < |b_{2k}| < |b_{2k+2}|$ $(k = 1, 2, \ldots)$. By (14), the absolute values of the numbers α_l^{-1} take on their minimum for $l = 1$ and the same is true for the numbers β_l^{-1}. By II, there is an $A = A_l$ with

$$f_a(\xi) = f_a(\eta)(b\eta + a)^{2q}, \qquad |b\eta + a| = 1.$$

If we put $\eta_A = \eta^*$, then η^* is also in $\mathfrak{F}$ and we have

$$f_a(\xi) = f_a(\eta^*).$$

If we write again η for η^*, then the assumption (13) remains satisfied and we now obtain the numbers $(b_l\eta + a_l)^{2q}$ $(l = 1, 2, \ldots)$ as a permutation of the numbers $(b_l\xi + a_l)^{2q}$. Hence there exists a permutation $1^*, 2^*, \ldots$ of the natural numbers $1, 2, \ldots$ and certain $2q$th roots of unity such that

$$b_l\xi + a_l = \varepsilon_l(b_{l^*}\eta + a_{l^*}) \qquad (l = 1, 2, \ldots). \tag{16}$$

Here the permutation in question and the choice of the roots of unity may depend on ξ and η. Since $\mathfrak{F}$ is contained in a disk $|z| < r < 1$, we have

$$|b_{l^*}| < |a_{l^*}|, \qquad |b_{l^*}\eta + a_{l^*}| > |a_{l^*}|\,(1 - r).$$

By (16), l^* is below a bound which depends only on l. A similar argument shows that for a given l^* the index l is bounded.

V. Let

$$b_k\xi + a_k = \varepsilon_k(b_{k^*}\eta + a_{k^*}) \tag{17}$$

be one of the equations (16). If

$$b_l\varepsilon_k b_{k^*} \neq b_k\varepsilon_l b_{l^*} \tag{18}$$

for given k and l, then ξ and η are uniquely determined as a solution of the two linear equations (16) and (17). Suppose $1^* \neq 1$. Let $l^* = 1$. The last sentence in IV above implies that the number of l with $l^* = 1$ is finite. Also, the number of $2q$th roots of unity is finite. Now it follows from (16) that the set of possible values for ξ is finite. We exclude these values for ξ.

Then $1^* = 1$ and $l^* > 1$ $(l = 2, 3, \ldots)$. Now let $l = 2, 3$ and $k^* = 2, 3$. We also exclude the finitely many possibilities for ξ which arise under the assumption (18) and so obtain

$$b_l \varepsilon_k b_{k^*} = b_k \varepsilon_l b_{l^*} \qquad (k^*, l = 2, 3). \tag{19}$$

Since $|b_2| = |b_3| < |b_k|$ $(k = 4, 5, \ldots)$, it follows from (19) that k and l^* must be 2 or 3. Hence either $2^* = 2$, $3^* = 3$ or $2^* = 3$, $3^* = 2$. With $k = 2$, $l = 3$, (16), (17), and (19) yield the relations

$$\varepsilon_2 = \varepsilon_3, \qquad b_2(a_3 - \varepsilon_3 a_{3^*}) = b_3(a_2 - \varepsilon_2 a_{2^*}),$$

so that

$$(1 - \varepsilon_2)(a_2 + \bar{a}_2) = 0.$$

Since A_2 is not elliptic, we have $a_2 + \bar{a}_2 \neq 0$ and $\varepsilon_2 = 1$. If we had $2^* = 2$, $3^* = 3$ then, by (17) with $k = 2$, it would follow that $\xi = \eta$, contradicting the assumed inequivalence of ξ and η. Hence $2^* = 3$, $3^* = 2$ and, now, (17) yields

$$b_2(\xi + \eta) = \bar{a}_2 - a_2. \tag{20}$$

VI. Since $\mathfrak{F}$ is compact, Γ is noncommutative. By Theorem 4 in Section 1, there is also an element $A = A_{2g}$ in Γ which does not commute with A_2, while A_{2k} for $k = 1, 2, \ldots, g - 1$ commutes with A_2. The reasoning carried out in the previous step for $k, l = 2, 3$ can be applied recursively for k, $l = 2n$, $2n + 1$ $(n = 2, \ldots, g)$. In this way obtain $(2n)^* = 2n + 1$, $(2n + 1)^* = 2n$, and

$$b(\xi + \eta) = \bar{a} - a; \tag{21}$$

here we put $b = b_{2g}$, $a = a_{2g}$. If a_2 were not real, then, by (20), $\xi + \eta$ would be different from 0 and we could put

$$\frac{-2i}{\xi + \eta} = \rho.$$

Then (20) and (21) would yield the parametric representation

$$a_2 = p_2 + iq_2, \qquad b_2 = \rho q_2, \qquad a = p + iq, \qquad b = \rho q$$

with real p_2, q_2, p, q. This would lead to the contradiction

$$AA_2 = \begin{pmatrix} \bar{a} & \bar{b} \\ b & a \end{pmatrix} \begin{pmatrix} \bar{a}_2 & \bar{b}_2 \\ b_2 & a_2 \end{pmatrix} = \begin{pmatrix} \bar{a}\bar{a}_2 + \bar{b}b_2 & \bar{a}\bar{b}_2 + \bar{b}a_2 \\ b\bar{a}_2 + ab_2 & b\bar{b}_2 + aa_2 \end{pmatrix} = A_2 A.$$

It follows that $a_2 = \bar{a}_2$, $\xi + \eta = 0$, $a = \bar{a}$. Since $AA_2 \neq A_2 A$, the second diagonal entry $b\bar{b}_2 + aa_2$ of AA_2 is not real. With AA_2 in place of A we obtain the previous contradiction. Here we note that in connection with the recursive steps from A_2 to A and AA_2 we must rule out finitely many additional values for ξ. This completes the proof.

In proving Theorem 3 we ruled out for ξ certain (finitely many) equivalence classes. The proof does not show that the theorem is invalid for these classes and nothing conclusive is known concerning this matter. The result just established, however, is sufficient for our purposes.

As a simple consequence of Theorem 3 we show that, conversely, the field K of automorphic functions associated with Γ determines the group Γ completely. We assume that (1) holds for a fixed mapping A in Ω and all the functions $f(z)$ in K and prove that A is in Γ. To this end we choose a point z_0 in $\mathfrak{E}$ which does not belong to any of the excluded equivalence classes and a sufficiently small neighborhood $\mathfrak{U}$ of z_0. By Theorem 3, every point ξ of $\mathfrak{U}$ is equivalent to ξ_A, that is, $\xi_A = \xi_B$, where B denotes an element of Γ which may depend on ξ. Since the number of admissible B is finite, there is a mapping B for which the equality above holds identically in ξ. This implies $A = B$, and our assertion follows.

In proving Theorem 3 we supposed Γ normalized. We shall show that we can dispense with this assumption. If C is an element of Ω, then the group $\Gamma^* = C\Gamma C^{-1}$ is conjugate to Γ in Ω and, clearly, $\mathfrak{F}_C$ is a fundamental region of Γ^*. Furthermore, the substitution $w = z_C$ corresponding to C carries every automorphic function $f(z)$ associated with Γ into the automorphic function $f(w_{C^{-1}})$ of w associated with Γ^*; this without the elements of the field K undergoing any change. We now show that for a suitable choice of C the group Γ^* is normalized, the only assumption being that Γ is a circle group. Here it will suffice to put

$$z_C = \frac{z - \zeta}{1 - \bar{\zeta} z} \qquad (|\zeta| < 1),$$

that is,

$$C = (1 - \zeta\bar{\zeta})^{-1/2}\begin{pmatrix} 1 & -\zeta \\ -\bar{\zeta} & 1 \end{pmatrix},$$

$$\begin{aligned}(1 - \zeta\bar{\zeta})CAC^{-1} &= \begin{pmatrix} 1 & -\zeta \\ -\bar{\zeta} & 1 \end{pmatrix}\begin{pmatrix} \bar{a} & \bar{b} \\ b & a \end{pmatrix}\begin{pmatrix} 1 & \zeta \\ \bar{\zeta} & 1 \end{pmatrix} \\ &= \begin{pmatrix} 1 & -\zeta \\ -\bar{\zeta} & 1 \end{pmatrix}\begin{pmatrix} \bar{a} + \bar{b}\bar{\zeta} & \bar{a}\zeta + \bar{b} \\ b + a\bar{\zeta} & b\zeta + a \end{pmatrix} \\ &= \begin{pmatrix} \bar{\alpha} & \bar{\beta} \\ \beta & \alpha \end{pmatrix},\end{aligned}$$

with

$$\beta = b + (a - \bar{a})\bar{\zeta} - \bar{b}\bar{\zeta}^2, \qquad a\bar{a} - b\bar{b} = 1.$$

If $\mathfrak{F}$ is a normal fundamental region of Γ with center ζ, then $\mathfrak{F}_C$ is a normal fundamental region of Γ^* with center 0. Now let

$$A_0 = \begin{pmatrix} \bar{a}_0 & \bar{b}_0 \\ b_0 & a_0 \end{pmatrix}, \qquad \beta_0 = b_0 + (a_0 - \bar{a}_0)\bar{\zeta} - \bar{b}_0\bar{\zeta}^2, \qquad a_0\bar{a}_0 - b_0\bar{b}_0 = 1.$$

If we have $|\beta| = |\beta_0|$ identically for a complex variable ζ, then

$$b_0 = \lambda b, \qquad a_0 - \bar{a}_0 = \lambda(a - \bar{a}), \qquad \bar{b}_0 = \lambda\bar{b},$$

with λ indendent of ζ and of absolute value 1. If $A \neq \pm E$ then $\lambda = \pm 1$, $a_0 - \lambda a = r$ real,

$$0 = a_0\bar{a}_0 - a\bar{a} = r^2 + r\lambda(a + \bar{a}),$$

so that $a_0 = \lambda a$ or $a_0 = -\lambda\bar{a}$. It follows that $A_0 = \pm A$ or $A_0 = \pm A^{-1}$. Now we let A, A_0 vary over the finite or countably infinite set of pairs of elements in Γ with $A_0 \neq \pm A$, $A_0 \neq \pm A^{-1}$ and determine a point ζ in $\mathfrak{E}$ such that all the corresponding inequalities $|\beta| \neq |\beta_0|$ hold. This is possible, for every equation $|\beta|^2 = |\beta_0|^2$ yields a curve of fourth order in the z-plane. For such a choice of ζ, Γ^* is normalized. It is an open question whether the transition from Γ to Γ^* is actually necessary for the proof of Theorem 3.

We note that the Poincaré series of special type formed for Γ^* can be rearranged by means of the relation

$$\sum_A \left(\frac{dw_{CAC^{-1}}}{dw}\right)^k = \left(\frac{dz_C}{dz}\right)^{-k} \sum_A \left(\frac{dz_{CA}}{dz}\right)^k \qquad (k = 2, 3, \ldots).$$

Theorem 3 implies, in particular, that among the functions $h_k(z)$ in (8) there is at least one which is not constant. We choose one such function and denote it by $f = f(z)$. Let f have order m and let Λ again be the field of rational functions of f with constant coefficients. Let K_0 be the field whose elements are rational functions with constant coefficients in finitely many of the functions $h_1(z)$, $h_2(z)$, . . . , where the denominators of the rational functions do not vanish identically in z. Then K_0 contains Λ and is contained in K. We now prove the important theorem which asserts that K_0 coincides with K. To this end we construct an element g in K_0 which when adjoined to Λ yields the field K.

Theorem 4: K_0 contains a function g whose m powers g^k $(k = 0, \ldots, m - 1)$ are linearly independent over Λ.

Proof: The proof is in three steps.

I. We study the mapping $w = f(z)$ from the normal polygon $\mathfrak{F}$ to the w-sphere. If we identify equivalent boundary points of $\mathfrak{F}$, then, by Theorem 1, the mapping in question covers the sphere m times. The mapping is schlicht at a point $z = \xi$ if and only if the value $w = f(\xi)$ is taken on to first order. Otherwise we obtain over the w-sphere a branch point of appropriate order so that, upon introduction of local parameters, the mapping is everywhere conformal. Joining equivalent sides of $\mathfrak{F}$ produced the closed surface $\mathfrak{F}^*$ which is now mapped onto an m-sheeted Riemann region $\mathfrak{R}$ over the w-sphere. The compactness of $\mathfrak{F}^*$ implies that $\mathfrak{R}$ is compact and so, by the result at the end of Section 8 of Chapter 2 (Volume I), is the Riemann surface of an algebraic function of w. We shall not make use of this result here. If we go from $\mathfrak{F}^*$ back to $\mathfrak{F}$ by opening the original edges of $\mathfrak{F}$, we induce a dissection of $\mathfrak{R}$. In this way we obtain over the w-sphere a simply connected surface $\underline{\mathfrak{R}}$ with boundary; that $\underline{\mathfrak{R}}$ is simply connected follows from the fact that it is a topological image of $\mathfrak{F}$ which, as a noneuclidean convex polygon, is simply connected. Here we note that the dissection of $\mathfrak{R}$ need not be canonical in the sense of Section 2 of Chapter 2 (Volume I).

II. By Theorem 2, K contains at most m automorphic functions which are linearly independent over Λ. It follows, in particular, that among the infinitely many functions $h_1(z), h_2(z), \ldots$ there is a maximal number s of linearly independent ones. If the s functions h_k $(k = n_1, \ldots, n_s)$ are linearly independent over Λ, then it is possible to choose constants $c_1, \ldots, c_s$ so that adjunction of the function

$$c_1 h_{n_1}(z) + \cdots + c_s h_{n_s}(z) = g(z) = g \tag{22}$$

to Λ yields the field K_0. This means that $K_0 = \Lambda(g)$ is a simple algebraic extension of degree s over Λ and g satisfies an irreducible equation over Λ,

$$g^s + \varphi_1 g^{s-1} + \cdots + \varphi_s = 0, \tag{23}$$

with coefficients $\varphi_1(z), \ldots, \varphi_s(z)$ in Λ. Since the powers g^k $(k = 0, \ldots, s-1)$ form a basis of K_0 over Λ,

$$h_k = \psi_{k1} g^{s-1} + \psi_{k2} g^{s-2} + \cdots + \psi_{ks} \qquad (k = 1, 2, \ldots), \tag{24}$$

where the coefficients $\psi_{k1}(z), \ldots, \psi_{ks}(z)$ are in Λ. Certainly, $s \leqslant m$. We shall show that $s = m$.

III. Let $w = \omega$ be a point on the complex w-sphere satisfying the following conditions. First, the points of the Riemann region $\mathfrak{R}$ over ω are not branch points, so that the equation $f(z) = \omega$ has exactly m distinct simple roots $z_1, \ldots, z_m$ on $\mathfrak{F}_0$. Second, $z_1, \ldots, z_m$ are different from the finitely many values excluded by the statement of Theorem 3. Third, $z_1, \ldots, z_m$ are different from the poles of the functions $\varphi_l(z)$ and $\psi_{kl}(z)$ $(l = 1, \ldots, s;\ k = 1, 2, \ldots)$. Our conditions bar only countably many of the uncountably many points on the w-sphere, and so there is an ω satisfying all of our requirements.

Since $\varphi_1, \ldots, \varphi_s$ are rational functions of f and $f(z)$ takes on the value ω at $z_1, \ldots, z_m$, it follows that the numbers $\varphi_1(z_k), \ldots, \varphi_s(z_k)$ are independent of the index $k = 1, \ldots, s$ and, by the third condition, finite. By (23), the m values $g(z_1), \ldots, g(z_m)$ are also finite and are roots of an algebraic equation of degree s. If we had $s < m$, then we could choose our subscripts so that $g(z_1) = g(z_2)$; here z_1 and z_2 are different points of $\mathfrak{F}_0$ and, therefore, inequivalent. But then (24) and the third condition would imply

$$h_k(z_1) = h_k(z_2) \qquad (k = 1, 2, \ldots),$$

which, using the second condition, would lead to a contradiction of Theorem 3. This completes the proof.

Theorem 4 implies directly that K_0 coincides with K, for K and its subfield K_0 have the same degree m over Λ. It follows that every automorphic function associated with Γ is a rational function of f and g with constant coefficients On the other hand, it is possible, using (8) and (22), to express every such rational function as the quotient of two automorphic forms of equal weight both of which are isobaric polynomials in the Poincaré series f_{kq} $(k = 1, 2, \ldots)$; in fact, it suffices to use the series corresponding to certain finitely many fixed indices. (We call a polynomial in automorphic forms with constant coefficients *isobaric* if all of its terms have equal weight.) This yields

Theorem 5: It is possible to choose finitely many Poincaré series of special type such that every automorphic function associated with Γ is expressible as a quotient of two isobaric polynomials of equal weight in these Poincaré series.

In the proof of Theorem 4 we made use of the assumption that the automorphic function f is not constant, but we made no use of the fact that f is of the form given in (8). If m is the order of a nonconstant automorphic function f associated with Γ, then there always exists another automorphic function belonging to Γ such that f and g are connected by an irreducible algebraic equation $P(f, g) = 0$ of degree m in g with constant coefficients. If, in this equation, we regard f as an independent variable, then g is an algebraic function of f.

Theorem 6: The Riemann surface of the algebraic function g of f is the image $\mathfrak{R}$ over the w-sphere, obtained via the function $w = f(z)$, of the fundamental region $\mathfrak{F}$ with equivalent boundary points identified.

Proof: If f takes on at a point ξ of $\mathfrak{F}$ the value $f(\xi) = \omega$ to order n, then we obtain for the inverse function $z = \varphi(w)$ in the vicinity of $w = \omega$ a series in powers of $t = (w - \omega)^{1/n}$, and t is then the local parameter at the point of $\mathfrak{R}$ over ω which corresponds to the point ξ of $\mathfrak{F}$; if $\omega = \infty$ we put $t = w^{-(1/n)}$. Since the function $g(z)$ is meromorphic on $\mathfrak{F}$ and takes on the same value at equivalent boundary points of $\mathfrak{F}$, it follows that, as a function of w, it is single-valued and meromorphic on the Riemann region $\mathfrak{R}$. The number of sheets of $\mathfrak{R}$ agrees with the degree m of the irreducible polynomial P in g, so that $\mathfrak{R}$ is indeed the Riemann surface of the algebraic function g of f.

The field K of automorphic functions associated with Γ consists of the rational functions in f and g. By Theorem 4 in Section 1 of the second chapter (Volume I), K consists of just those functions $h(z)$ which, as functions of $w = f(z)$, are meromorphic on $\mathfrak{R}$. Here we note that the Riemann surface $\mathfrak{R}$ depends on the choice of the function $f(z)$ for which we may take any nonconstant function in K. To point up this dependence we write $\mathfrak{R}(f)$ in place of $\mathfrak{R}$.

Theorem 7: All surfaces $\mathfrak{R}(f)$ can be mapped conformally onto each other and have the same genus.

Proof: Every Riemann surface $\mathfrak{R}(f)$ is the conformal image of the fundamental region $\mathfrak{F}$ with equivalent boundary points identified. Since $\mathfrak{F}$ is independent of the choice of the function f, the first part of our assertion follows.

If two compact Riemann surfaces $\mathfrak{R}$ and $\mathfrak{R}^*$ are mapped conformally onto each other, then this mapping induces an isomorphic correspondence between their fundamental groups. By Theorem 2 in Section 4 of the second chapter (Volume I), $\mathfrak{R}$ and $\mathfrak{R}^*$ have the same genus. We could also use (18) in Section 4 and argue as follows. We take as known the result from topology that the genus, computed by means of the Euler-Poincaré formula, is independent of the triangulation of the manifold. Then the quantity denoted by p in (18), Section 4, is actually the genus of $\mathfrak{R}(f)$ in the sense of our earlier definition and obviously independent of f. At the same time we see that $p > 1$, a conclusion which we will soon establish by means of a different argument.

Theorem 8: Let $\mathfrak{U}(f)$ be the covering surface of the Riemann surface $\mathfrak{R}(f)$. Then the function $w = f(z)$ maps the interior of the unit disk in the z-plane conformally onto $\mathfrak{U}(f)$.

Proof: Let $\mathfrak{F}_A$ be the image of the normal polygon $\mathfrak{F}$ under the mapping A in Γ. The function $w = f(z)$ carries every region $\mathfrak{F}_A$ into the same image region $\underline{\mathfrak{R}}$ which is obtained from $\mathfrak{R}(f)$ by dissection along the image of the boundary of $\mathfrak{F}$. By associating with each $\mathfrak{F}_A$ a copy of $\underline{\mathfrak{R}}$ and by joining the latter in the manner of the $\mathfrak{F}_A$ in the z-plane, we obtain a Riemann region $\mathfrak{B}$ over the w-sphere. By our construction, $\mathfrak{B}$ is the conformal image, under $w = f(z)$, of the unit disk $\mathfrak{E}$, and we must now show that $\mathfrak{B}$ coincides with $\mathfrak{U}(f)$. To begin with, $\mathfrak{B}$ and $\mathfrak{U}(f)$ share the region $\underline{\mathfrak{R}}$ in which we choose a point $\mathfrak{p}$. To every curve Q on $\mathfrak{R}$ issuing from $\mathfrak{p}$ there corresponds a unique curve L, on $\mathfrak{B}$, whose projection is Q. In particular, let Q be a closed curve on $\mathfrak{R}$. We must show that L is closed on $\mathfrak{B}$ if and only if Q is homotopic to zero on $\mathfrak{R}$. The mapping effected by means of the function $w = f(z)$ enables us to transplant the curve L in a one-to-one manner to a curve M on $\mathfrak{E}$ such that $\mathfrak{p}$ corresponds to the initial point of M.

If the curve Q is homotopic to zero then we can contract it to $\mathfrak{p}$. Since the terminal point of M also remains fixed in the course of this deformation process, M and L are both closed. The converse conclusion is equally obvious owing to the simple-connectedness of $\mathfrak{E}$. This completes the proof.

In view of Theorem 1 of Section 9 in the second chapter (Volume I), Theorem 8 implies that $p > 1$; this is so because $\mathfrak{U}(f)$ is hyperbolic.

Another consequence of Theorem 8 is that the inverse function $z = \varphi(w)$ of $w = f(z)$ is analytic on the covering surface $\mathfrak{U}$. Since $\mathfrak{E}$ is schlicht, $\varphi(w)$ takes on different values at different points of $\mathfrak{U}$, so that $\mathfrak{U}$ is the Riemann surface of the inverse $\varphi(w)$ of $f(z)$. The conformal mapping $z = \varphi(w)$ carries $\mathfrak{U}$ into $\mathfrak{E}$. If z is viewed as a function of z, then the associated Riemann surface is not $\mathfrak{E}$ but the z-sphere. This observation shows that under conformal mapping of the Riemann surface of an analytic function the image surface need not be the Riemann surface of the transformed function. This is why the now prevalent abstract definition of a Riemann surface is hardly commendable.

Theorem 9: The fundamental group of $\mathfrak{R}$ is isomorphic to Γ.

Proof: Let ξ be the point of $\mathfrak{F}$ corresponding to $\mathfrak{p}$. With L the same as in the proof of Theorem 8, the terminal point of L lies at the point $\mathfrak{p}$ of a copy of $\underline{\mathfrak{R}}$ which is the image of a definite region $\mathfrak{F}_A$, so that ξ_A is the terminal point of M. Since Γ is hyperbolic, $\xi_A = \xi$ if and only if A is the identity mapping. If we keep $\mathfrak{p}$ fixed and deform Q, then the terminal point of M likewise remains fixed. It follows that A depends only on the homotopy class of Q. Conversely, let A be an element of Γ. By joining ξ to ξ_A within $\mathfrak{E}$ we obtain a curve M and from it again a closed curve Q on $\mathfrak{R}$ issuing from $\mathfrak{p}$. In this way we establish a one-to-one correspondence between the mappings A in Γ and the elements A^* of the fundamental group of $\mathfrak{R}$.

Now, if A and B are two elements in Γ and $AB = C$, then we obtain a curve in $\mathfrak{E}$ going from ξ to ξ_C by first joining ξ to ξ_A, and then adjoining to it the image under A of a curve from ξ to ξ_B. Using the mapping $w = f(z)$, we obtain immediately the relation $A^*B^* = C^*$. This proves our theorem.

7. *Automorphic and algebraic functions*

In the preceding section we started with the field of automorphic functions associated with a hyperbolic polygon group Γ and studied the corresponding

Riemann surfaces $\mathfrak{R}(f)$. The existence of hyperbolic polygon groups remained an open question.

Now we start, conversely, with a compact Riemann region $\mathfrak{R}$ over the w-sphere. We showed in Section 9 of Chapter 2 (Volume I) that there exists an algebraic function $g = g(w)$ whose Riemann surface is $\mathfrak{R}$. Now let

$$P(g, w) = 0 \tag{1}$$

be the irreducible equation for g of degree n in g. The meromorphic functions on $\mathfrak{R}$ yield a function field which we denote by K. The field K consists of all rational functions $R(g, w)$ in g and w in which the denominator does not contain, identically in g and w, the factor $P(g, w)$. By Theorem 4 in Section 1 of Chapter 2 (Volume I), the elements of K can be described more precisely as polynomials in g of degree less than n whose coefficients are rational functions of w. Thus (1) defines g as an algebraic function of the independent variable w.

If

$$g^* = R_1(g, w), \qquad w^* = R_2(g, w) \tag{2}$$

are two functions in K of which w^* is not constant, then elimination of g and w with the aid of (1) yields an irreducible algebraic equation

$$Q(g^*, w^*) = 0 \tag{3}$$

which actually contains g^*. It is clear that the algebraic function field K* generated by g^* and w^* is contained in K. The field K* coincides with K if and only if it is possible to express g and w rationally in terms of g^* and w^*, that is, if and only if

$$g = R_3(g^*, w^*), \qquad w = R_4(g^*, w^*). \tag{4}$$

The equations (2) can be viewed as a *rational transformation* of coordinates which carries the algebraic curve defined by (1) into the algebraic curve defined by (3). If, in addition, (4) holds, then we speak of a *birational transformation* of the curves into each other. In the latter case, (2) need not be a *Cremona transformation*, that is, a birational transformation of the two independent variables g and w.

Two questions suggest themselves. One question is: given a nonconstant function $w^* = R_2(g, w)$, is it always possible to find another function $g^* = R_1(g, w)$ such that K* coincides with K? The other question is: with $\mathfrak{R}^*$ the Riemann surface of g^* over the w^*-sphere, what is the connection between the Riemann surfaces $\mathfrak{R}$ and $\mathfrak{R}^*$? These questions have already been answered, in appropriate formulation, for the fields of automorphic functions by Theorems 4 and 7 of the preceding section, and we now discuss them from our present point of view.

Theorem 1: If w^* is an arbitrary nonconstant function in K, then there exists a corresponding function g^* such that the field generated by g^* and w^* coincides with K. Furthermore, the correspondence from w to w^* maps the Riemann surface $\mathfrak{R}$ conformally onto the Riemann surface $\mathfrak{R}^*$.

Proof: We show, first of all, that for an arbitrary choice of g^* in K the degree, in g^*, of the polynomial Q in (3) lies below a bound which is independent of g^*. In fact, if we view R_1 and R_2 as polynomials in g whose coefficients are rational functions of w then, by forming the resultants relative to g, we obtain from the two pairs of equations

$$P(g, w) = 0, \quad R_1(g, w) - g^* = 0; \quad P(g, w) = 0, \quad R_2(g, w) - w^* = 0$$

two algebraic equations

$$A_1(g^*, w) = 0, \qquad A_2(w^*, w) = 0,$$

of which the first is of degree n in g^* and the second is independent of the choice of g^*. By forming the resultant of these equations with respect to w, we obtain an algebraic equation for g^* depending on w^* whose degree lies below a bound which is independent of g^*. The analogous statement is then all the more true of the irreducible equation (3). This proves, in particular, that K is an algebraic extension of the field of rational functions of w^*, and so a simple algebraic extension. By choosing g^* so that Q is of highest possible degree in g^*, we obtain a proof of the first part of the theorem.

Let $\mathfrak{p}$ be a point of the Riemann surface $\mathfrak{R}$, possibly a branch point, and t the corresponding local parameter. Then $g^* = g^*(t)$ and $w^* = w^*(t)$ are meromorphic functions of t in the vicinity of $\mathfrak{p}$. For $t \neq 0$ the values $g^*(t)$ and $w^*(t)$ determine uniquely, in a sufficiently small neighborhood, a point on the Riemann surface $\mathfrak{R}^*$ which, for $t \to 0$, goes to (say) $\mathfrak{p}^*$. The corresponding local parameter t^* is a single-valued regular function of t in the deleted neighborhood of $t = 0$ which vanishes at $t = 0$. An analogous statement holds for t viewed as a function of t^*. This implies the power series expansion

$$t^* = c_1 t + c_2 t^2 + \cdots,$$

with $c_1 \neq 0$. It follows that the neighborhood of $t = 0$ is mapped conformally onto the neighborhood of $t^* = 0$, and the rest of our theorem is proved.

A direct consequence of Theorem 1 is that the fundamental groups of $\mathfrak{R}$ and $\mathfrak{R}^*$ are isomorphic, so that both surfaces have the same genus p. This justifies calling p the *genus of the field* K and, at the same time, the *genus of the algebraic curve* defined by (1); the genus, then, is invariant under birational transformations.

Now we make use of the theorems on uniformization established in Section 10 of Chapter 2 (Volume I). These theorems assert that it is possible to determine a function $z = \varphi(w)$ which maps the covering surface $\mathfrak{U}$ of the compact Riemann surface $\mathfrak{R}$ conformally onto the normal region $\mathfrak{S}$, where $\mathfrak{S}$ is the z-sphere for $p = 0$, the z-plane for $p = 1$, and the unit disk $\mathfrak{E}$ for $p > 1$. Also, the covering transformations of $\mathfrak{U}$ yield a faithful representation of the fundamental group of $\mathfrak{R}$ by means of a group Γ of conformal mappings of $\mathfrak{S}$ onto itself which have no fixed points. The following theorem was proved at the end of Chapter 2 (Volume I).

Theorem 2: If $p = 1$ then Γ is a group of translations $s = z + ka + lb$ $(k, l = 0, \pm 1, \ldots)$ in the plane of complex numbers and the ratio of the two constants a and b is not real.

Theorem 3: If $p > 1$ then Γ is a hyperbolic polygon group.

Proof: The elements of Γ are conformal mappings of $\mathfrak{E}$ onto itself and so are of the form

$$s = \frac{\bar{a}z + \bar{b}}{bz + a} \qquad (a\bar{a} - b\bar{b} = 1). \tag{5}$$

When these mappings are different from the identity, they have no fixed points in $\mathfrak{E}$, and so are hyperbolic or parabolic.

The covering surface $\mathfrak{U}$ was constructed in Section 3 of Chapter 2 (Volume I) by joining copies $\mathfrak{F}_\alpha$ of the canonically dissected Riemann region $\mathfrak{F}$, where α varies over the fundamental group of the nondissected Riemann surface $\mathfrak{R}$. Since $\mathfrak{F}$ is compact on $\mathfrak{U}$, the image $\mathfrak{B}$ of $\mathfrak{F}$ under the function $z = \varphi(w)$ is compact in $\mathfrak{E}$. Since, in connection with the dissection of $\mathfrak{R}$ leading to $\mathfrak{F}$, the cut curves are supposed piecewise smooth, the boundary of $\mathfrak{B}$ is piecewise smooth. If A is the element of Γ corresponding to α, then $\mathfrak{B}_A$ is the image of $\mathfrak{F}_\alpha$ under the mapping of $\mathfrak{U}$ onto $\mathfrak{E}$, and, since this mapping is, in particular, one-to-one and onto, the domains $\mathfrak{B}_A$ yield a simple covering of $\mathfrak{E}$ without gaps. Hence $\mathfrak{B}$ is a fundamental region of Γ. Since $\mathfrak{B}$ is compact in $\mathfrak{E}$ it follows, by Theorem 4 in Section 4, that every normal fundamental region of Γ is compact. By Theorem 3 in Section 4, Γ contains no parabolic elements. This means that Γ is a hyperbolic polygon group as asserted.

Theorem 3 proves, in particular, the existence of a hyperbolic polygon group and, better still, the existence of infinitely many such groups, for, according to a remark in Section 4 of Chapter 2 (Volume I), there are compact Riemann regions of arbitrary genus.

In the case $p > 1$, the inverse $w = f(z)$ of the function $z = \varphi(w)$ is an automorphic function associated with Γ. If m is the number of sheets of the Riemann surface $\mathfrak{R}$, then, in view of the conformal mapping of $\mathfrak{B}$ onto $\mathfrak{F}$, the function $f(z)$ takes on every value m times in the fundamental region, so that $\mathfrak{O}(f) = m$. Further, if $g = g(w)$ denotes any meromorphic function on $\mathfrak{R}$, that is, a function in K, then the composite function $g(f(z))$ is an automorphic function associated with Γ. Conversely, for every automorphic function $h(z)$ associated with Γ, the function $h(\varphi(w))$ is single-valued and meromorphic on $\mathfrak{R}$. This means that, as a result of the introduction of the uniformizing parameter z, the algebraic function field K goes over into the field of automorphic functions associated with Γ. On the other hand, the results of the previous section show that all the hyperbolic polygon groups and the corresponding fields of automorphic functions arise in this way. If we uniformize a given Riemann region $\mathfrak{R}$ of genus $p > 1$ and then construct the group Γ, we obtain, by forming appropriate Poincaré series, explicit expressions for the functions of the corresponding field K. In particular, this gives another proof of the existence of algebraic functions with $\mathfrak{R}$ as Riemann surface. Here it must be borne in mind that the mapping function $z = \varphi(w)$ was obtained in the second chapter (Volume I) using the Dirichlet principle which does not automatically yield a constructive procedure for actual computation.

Since the most general conformal orientation-preserving mapping of $\mathfrak{E}$ onto itself is given by a linear fractional transformation of the form (5), that is, by an element V of the group Ω, it follows that the uniformizing parameter z is determined up to such a fixed transformation. Upon application of this transformation, Γ is replaced by the group $V\Gamma V^{-1}$, conjugate to Γ in Ω. On the other hand, by Theorem 1, if w is replaced by w^* and, correspondingly, $\mathfrak{R}$ by $\mathfrak{R}^*$, then the same uniformizing parameter z is admissible. This means that Γ is determined up to an arbitrary conjugate transformation V by K alone.

In the proof of Theorem 3 we made use of the fact that the covering surface $\mathfrak{U}$ is mapped conformally onto the normal region $\mathfrak{E}$, but not of the fact that then, by Theorem 1 in Section 10 of the second chapter, the genus p is >1. We now give a new proof of this result.

Since it is the conformal image of the canonically dissected Riemann surface, the fundamental region $\mathfrak{B}$ is a (not necessarily noneuclidean) polygon in $\mathfrak{E}$ with $4p$ vertices bounded by $4p$ smooth curves. We wish to show that for the noneuclidean area of $\mathfrak{B}$ we have

$$I(\mathfrak{B}) = 4(p - 1)\pi. \tag{6}$$

If we can establish this result, then, in view of $I(\mathfrak{B}) > 0$, it will follow that $p > 1$.

To simplify computations we replace $\mathfrak{E}$ with the model $\mathfrak{H}$ of the noneuclidean plane. In analogy with (13) in Section 2, we obtain by Green's formula the relation

$$I(\mathfrak{B}) = \iint\limits_{\mathfrak{B}} \frac{dx\,dy}{y^2} = \int_C \frac{dx}{y} \qquad (z = x + iy),$$

where the boundary C is positively oriented. The substitution $x = (z + \bar{z})/2$, $y = (z - \bar{z})/2i$ yields

$$I(\mathfrak{B}) = i\int_C \frac{dz}{z - \bar{z}} - i\int_C \frac{d\bar{z}}{\bar{z} - z}. \tag{7}$$

Let S, S' be a pair of equivalent sides of $\mathfrak{B}$, where S has the same orientation as C, and S' has the opposite orientation. Furthermore, let

$$z' = \frac{az + b}{cz + d}$$

be the fractional linear substitution of $\mathfrak{H}$ onto itself which carries S into S', where a, b, c, d are real and $ad - bc = 1$. Then

$$\begin{aligned} z' - \bar{z}' &= \frac{z - \bar{z}}{(cz + d)(c\bar{z} + d)}, \\ dz' &= \frac{dz}{(cz + d)^2}, \end{aligned} \tag{8}$$

so that

$$\frac{dz'}{z' - \bar{z}'} = \frac{c\bar{z} + d}{cz + d}\,\frac{dz}{z - \bar{z}},$$

$$\frac{dz}{z - \bar{z}} - \frac{dz'}{z' - \bar{z}'} = \left(1 - \frac{c\bar{z} + d}{cz + d}\right)\frac{dz}{z - \bar{z}} = \frac{c}{cz + d}\,dz = d\log(cz + d). \tag{9}$$

We introduce polar coordinates and put $cz + d = re^{i\varphi}$. Then, by (7) and (8), we see that the two sides S and S' of $\mathfrak{B}$ contribute to $I(\mathfrak{B})$ an amount equal to the change in the angle -2φ over S. Since, on the other hand, thc tangent vector on S is given by dz, (8) shows that -2φ is the angle between the direction of the tangent at the point z of S and the direction of the tangent at the corresponding point z' of S'. It follows that the change in this angle over S is equal to the difference of the changes in the directions of the tangents over S' and S. Noting orientations and considering all the pairs S, S' of C we find that $-I(\mathfrak{B})$ is equal to the sum of the $4p$ direction changes to which the tangent in traversing C is subject on the individual sides of the polygon. Now, at a vertex of $\mathfrak{B}$ the direction of the tangent jumps by $\pi - \omega$, where ω is the angle of $\mathfrak{B}$ at the vertex in question. Since C is a simple closed curve, the total change in the direction of the tangent corresponding to a single traversal in the positive direction is 2π. Hence

$$-I(\mathfrak{B}) + \sum_{\omega} (\pi - \omega) = 2\pi. \tag{10}$$

Here we made repeated use of the fact that the $4p$ sides are smooth curves. If we identify equivalent boundary points of $\mathfrak{B}$, then all the $4p$ vertices coalesce into a single point, as in the transition from $\mathfrak{F}$ to $\mathfrak{R}$. Hence the full neighborhood of a vertex ξ of $\mathfrak{B}$ is filled in a schlicht manner by $4p$ images $\mathfrak{B}_A$ where A varies over the elements of Γ which carry the $4p$ vertices into ξ. But then the $4p$ angles with vertex ξ coincide with the $4p$ individual angles ω at the vertices and their sum is 2π. It follows that

$$\sum_{\omega} (\pi - \omega) = 4p\pi - 2\pi.$$

By (10) this implies the assertion (6).

In conjunction with Theorem 5 of Section 4, (6) implies that p is independent of the canonical dissection of $\mathfrak{R}$. This was proved differently in Section 4 of the second chapter (Volume I).

With the aid of Theorem 5, Section 4, we obtain from (6) the formula (18) in Section 4, with p denoting the genus of $\mathfrak{R}$ as introduced in the original definition in Section 4; in the formula in question, p was obtained by application of the formula of Euler and Poincaré to the polyhedron $\mathfrak{F}^*$. We have not given a general proof of the topological theorem which asserts the equality of the two differently defined numbers, but the equalities (18), Section 4, and (6) constitute a proof of this theorem for the case under consideration.

8. Algebraic curves of genus 0 and 1

We now try to carry over the results established in the preceding section for $p > 1$ to the special cases $p = 1$ and $p = 0$ and to uncover their peculiar features.

In the case $p = 1$, the normal region $\mathfrak{S}$ is the (finite) z-plane. By Theorem 2, Section 7, the role of the automorphic functions is taken by the field K of elliptic functions with the two given periods a and b whose ratio is not real. It follows that we can apply the theorems of Section 14 of the first chapter (Volume I). We put, accordingly, $a = \omega_1$, $b = \omega_2$.

Clearly, the group Γ of translations

$$s = z + n_1\omega_1 + n_2\omega_2 \qquad (n_1, n_2 = 0, \pm 1, \ldots) \tag{1}$$

is discontinuous in the complex plane and has the parallelogram $\mathfrak{P}$ on ω_1 and ω_2 as fundamental region. By Theorem 3 in Section 14 of the first chapter (Volume I), every nonconstant function $w = f(z)$ in K assumes on $\mathfrak{P}$ every preassigned value the same number m of times; m was called the *order of the elliptic function*. This is analogous to Theorem 1 in Section 6. Furthermore, it is possible to show, in analogy to Theorem 1 of Section 6, that every $m + 1$ elliptic functions are linearly dependent over the field Λ of rational functions of w and to conclude that K arises out of Λ by adjunction of a suitable elliptic function $g(z)$ which is connected with $f(z)$ by means of an algebraic equation. The algebraic aspects of K can also be deduced from Theorem 6 of Section 14, Chapter 1 (Volume I), if we make use of Theorem 1 of Section 7.

In Theorem 3 of Section 6 the properties of Poincaré series of special type played a role, and with their aid we constructed the field of automorphic functions for a given polygon group Γ. Since for the group of translations defined by (1) the derivative of s with respect to z is 1, we cannot use the sums constructed in exact analogy to the Poincaré series. A reasonable alternative is to start with the expansions of $w = \wp(z)$ and its derivatives in series of partial fractions. Putting $\omega = n_1\omega_1 + n_2\omega_2$ we have, by Section 13 of Chapter 1 (Volume I), the representations

$$\wp(z) = z^{-2} + \sum_{\omega \neq 0} ((z + \omega)^{-2} - \omega^{-2}),$$

$$(2) \qquad \wp'(z) = \frac{d\wp(z)}{dz} = -2 \sum_{\omega} (z + \omega)^{-3}.$$

This gives directly elliptic functions with poles, whereas before we used the Poincaré series to construct automorphic forms whose quotients yielded automorphic functions. In view of (2), it is clear that $\wp(z)$ is independent of the choice of the basis ω_1, ω_2 of the period lattice, that is, of the choice of generators of Γ. Every other basis is of the form $\alpha\omega_1 + \beta\omega_2$, $\gamma\omega_1 + \delta\omega_2$, with integers α, β, γ, δ and $\alpha\delta - \beta\gamma = \pm 1$; here we take the plus sign if the new period parallelogram has the same orientation as the old one. This was proved in Section 9 of the first chapter (Volume I), where we also showed, notwithstanding differences of terminology, that every group of translations discontinuous in the complex plane is an abelian group with at most two independent generators. Corresponding to the three possibilities we have as fundamental domains the plane, a parallel strip or a parallelogram, and it is precisely in the last case that the fundamental region has finite euclidean area. Then we suppose $\mathfrak{P}$ fixed and given.

The relations (2) define on $\mathfrak{P}$ a nonconstant function $w = f(z) = \wp(z)$ of order $m = 2$. The $m + 1$ functions $(d\wp/dz)^2$, $d\wp/dz$, and 1 are linearly dependent over the field Λ of rational functions of w; not so $d\wp/dz$ and 1, since $\wp(z)$ is even and $d\wp(z)/dz$ is odd. For the same reason the quadratic equation for $d\wp/dz$ contains no linear term so that $(d\wp/dz)^2$ must be a rational function of w. Specifically, by Section 13, Chapter 1 (Volume I), we have

$$(3) \qquad \left(\frac{d\wp}{dz}\right)^2 = 4\wp^3 - g_2\wp - g_3, \qquad g_2 = 60 \sum_{\omega \neq 0} \omega^{-4}, \qquad g_3 = 140 \sum_{\omega \neq 0} \omega^{-6}.$$

It is now easy to carry over the Theorems 3 through 9, Section 6, in a natural way to the case under consideration except that now in place of the automorphic functions h_k $(k = 1, 2, \ldots)$ we have the functions $\wp$ and $\wp'$. If we used solely the idea of Theorem 3 just mentioned, then, to begin with, there would enter derivatives of $\wp(z)$ of arbitrarily high order, and these could then be eliminated using the differential equation

$$\frac{d^2\wp}{dz^2} = 6\wp^2 - \frac{g_2}{2}$$

which follows from (3). The following is a simpler argument. Let ξ and η be two points in the complex plane and $\xi - \eta$ not a period ω. Since $\wp(z)$ has order 2 and is even, the assumption $\wp(\xi) = \wp(\eta)$ implies that ξ is not a period and $\eta = \omega - \xi$ is a period. It follows that $\wp'(\eta) = -\wp'(\xi)$. If we also had $\wp'(\xi) = \wp'(\eta)$, then we would have $\wp'(\xi) = 0$, which is impossible, for by the assumption on $\xi - \eta$ the value $\wp(\xi)$ is taken on precisely to the first order. Hence the values of the two functions $\wp(z)$ and $\wp'(z)$ separate the classes of inequivalent points z. This is the required analog of Theorem 3 of Section 6. Furthermore, by Theorem 6, Section 14, Chapter 1 (Volume I), every elliptic function can be expressed rationally in terms of $\wp$ and $\wp'$; this is the analog of Theorem 5, Section 6.

Putting $f = \wp$, $h = \wp'$ we obtain, in analogy to Theorem 6, Section 6, the assertion that $w = \wp(z)$ maps the fundamental region $\mathfrak{P}$ conformally onto the canonically dissected two-sheeted Riemann surface of the algebraic function h in f defined by the equation

$$h^2 = 4f^3 - g_2 f - g_3. \tag{4}$$

More generally, if we take for f any nonconstant elliptic function, then the field K is obtained trivially by simultaneous adjunction of $\wp$ and $\wp'$ and, therefore, also by the simple adjunction of $g = \lambda\wp + \mu\wp'$ with suitable constants λ and μ. If f has order m, then, in accordance with Theorem 4, Section 6, the irreducible equation in f and g is of degree m in g. Now the remaining theorems of Section 6 carry over directly. In particular, we obtain, as a special case of the analog of Theorem 8, Section 6, the result that $w = \wp(z)$ maps the z-plane conformally onto the covering surface of the two-sheeted Riemann surface defined by (4). This was proved in a different form in the first chapter (Volume I) in Theorem 2, Section 8, in conjunction with Section 13.

By Section 10 of Chapter 2 (Volume I), the most general conformal and orientation-preserving mapping of the z-plane onto itself is given by a linear transformation $s = \alpha z + \beta$, with arbitrary constant $\alpha \neq 0$ and β, which we denote by V. Under the euclidean similarity transformation V, the elements $s = z + \omega$ of the group of translations Γ go over into the elements $s = z + \alpha^{-1}\omega$ of $V\Gamma V^{-1}$, so that we can normalize the fundamental periods ω_1, ω_2 and put $\omega_1 = 1$. Conversely, given a discontinuous group of translations Γ with compact fundamental region, there is exactly one field of elliptic functions which is algebraic with respect to these functions and of genus $p = 1$.

In the case $p > 1$, construction of a field of automorphic functions required the use of the Poincaré series. While very important, Theorems 3 and 4, Section 6, give us no truly usable procedure for the construction of two generating functions f and g connected by an algebraic equation of a certain normal form, as exemplified by (4) for $p = 1$. For this very case it is natural to ask to what extent the field K determines the coefficients g_2, g_3.

Theorem 1: Every algebraic curve of genus 1 can be transformed birationally into a cubic curve of the special form

$$h^2 = 4f^3 - g_2 f - g_3 \qquad (g_2^2 - 27g_3^2 \neq 0) \tag{5}$$

with constants g_2, g_3. Two such cubic curves are birationally equivalent if and only if they agree on the invariant

$$j = \frac{g_2^3}{g_2^3 - 27g_3^2}. \tag{6}$$

If this is the case, then the two curves go over into each other under an affine transformation of the form $f \to \alpha^2 f$, $h \to \alpha^3 h$, with constant $\alpha \neq 0$.

Proof: The existence of the normal form (5) follows from the fact that the algebraic function field of genus 1 defined by the curve is uniformized by elliptic functions $h = \wp'(z)$, $g = \wp(z)$ connected by the equation (3). The inequality between g_2 and g_3 in (5) was deduced in Section 13 of the first chapter (Volume I).

Since two algebraic curves are *birationally equivalent* if and only if they define the same field K, and in the case $p = 1$ the field K determines the group of translations Γ up to a euclidean similarity transformation, it follows that the periods ω of the $\wp$ – function which effects the uniformization of the first curve go over into the corresponding periods of the second curve under a transformation $\omega \to \alpha\omega$ with constant $\alpha \neq 0$. In view of (3), g_2 and g_3 are multiplied by factors α^{-4} and α^{-6} and, in view of (6), j remains invariant. Conversely, prescribing j determines g_2 and g_3 up to such factors with arbitrary $\alpha \neq 0$. The transformation $f \to \alpha^2 f$, $h \to \alpha^3 h$ yields in (5) just those factors for g_2 and g_3. This completes the proof.

It remains to discuss the case $p = 0$. In that case $\mathfrak{S}$ is the z-sphere and the fundamental group consists of the identity alone, so that no special invariance requirement must be met by the functions $w = f(z)$ which we are about to investigate. The only meromorphic functions on the sphere are the rational functions, and K is therefore the field of rational functions of z. We can again carry over the theorems of the last but one section with appropriate modification and partly trivial proofs.

Every rational function is of the form $w = P(z)/Q(z)$ with relatively prime polynomials $P(z)$ and $Q(z)$ of which Q is not identically 0. The function w is constant if and only if P and Q are constant. If this case is excluded, then the irreducible algebraic equation

$$wQ(z) - P(z) = 0 \tag{7}$$

is of positive degree m in z, where m is the maximum of the degrees of P and Q; we call m the *degree* of w. The relation (7) defines z as an algebraic function of w whose Riemann surface over the w-sphere is $\mathfrak{R}$. Since $\mathfrak{R}$ has m sheets, the function $w = f(z)$, viewed as a function of z on the z-sphere, takes on every value (counting multiplicities) m times. This result is again analogous to Theorem 1, Section 6. Furthermore, if Λ denotes the field of rational functions of w, then K is algebraic of degree m over Λ, and every $m + 1$ functions in K are again linearly dependent over Λ, in analogy to Theorem 2, Section 6. In place of the automorphic functions $h_k(z)$ $(k = 1, 2, \ldots)$ for $p > 1$ and the elliptic functions $\wp(z)$, $\wp'(z)$ for $p = 1$, we now have the function z itself. We note, too, that $w = f(z)$ maps the z-sphere conformally onto $\mathfrak{R}$, and this is the analog of Theorem 6 in Section 6. No further new results correspond to the remaining theorems in Section 6.

Conversely, let $\mathfrak{R}$ be a compact Riemann region of genus 0. $\mathfrak{R}$ coincides with its covering surface. If z is the uniformizing parameter, then the sphere is mapped conformally onto $\mathfrak{R}$ by a meromorphic function $w = f(z)$, which again shows w to be a rational function of z. It follows that all Riemann surfaces of genus 0 arise as rational mappings $w = f(z)$ of the z-sphere.

The algebraic curve given by (7) admits the trivial parametrization $z = t$, $w = P(t)/Q(t)$, where t is seen to be the independent complex parameter. We now answer the natural question, namely, how one obtains all the irreducible algebraic curves which have a rational parametric representation in an independent variable t. In other words, if

$$A(u, v) = 0 \tag{8}$$

is the irreducible equation of the curve, then we require two rational functions

$$u = \varphi(t), \qquad v = \psi(t) \tag{9}$$

of a variable t, not both constant, which satisfy the equation (8) identically in t. An algebraic curve of this type is said to be a *rational algebraic curve.*

Theorem 2: An algebraic curve is rational if and only if it is of genus $p = 0$.

Proof: At least one of the variables u and v, say v, actually appears in (8). Let $\mathfrak{R}$ be the Riemann surface of v as a function of u. If the genus is 0,

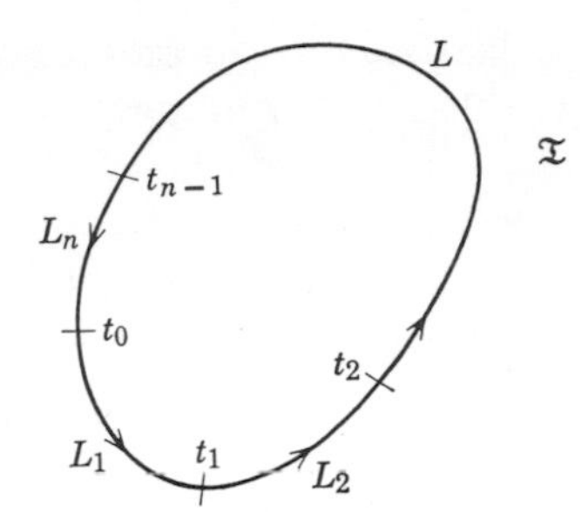

Figure 31

then u and v are rational functions of the uniformizing parameter t, so that (9) holds.

Now we must show that, conversely, the existence of the rational parametric representation (9) implies that the genus p of the curve (8) is 0, and this is more difficult. Let $\varphi(t)$ be nonconstant. As t varies over the complex sphere $\mathfrak{T}$, the latter is mapped uniquely by (9) onto a portion $\mathfrak{R}_0$ of $\mathfrak{R}$. Introduction of local parameters on $\mathfrak{R}$ shows that $\mathfrak{R}_0$ is open. On the other hand, since $\mathfrak{T}$ and $\mathfrak{R}$ are compact, it follows that $\mathfrak{R}_0$ is compact, so that $\mathfrak{R}_0$ coincides with $\mathfrak{R}$. If we exclude from $\mathfrak{T}$ and, correspondingly, from $\mathfrak{R}$, the finitely many points $t = \tau$ at which $t = \infty$, or $\varphi(t) = \infty$, or $\varphi'(t) = 0$, then the mapping of $\mathfrak{T}$ onto $\mathfrak{R}$ is locally one-to-one. However, the mapping need not be one-to-one in the large, so that, in general, a point of $\mathfrak{R}$ is the image of more than one point of $\mathfrak{T}$.

Let t_0 be different from the points τ. Beginning with the image of t_0 on $\mathfrak{R}$, we traverse on $\mathfrak{R}$ a closed curve C which bypasses the images of the points τ. This curve is the image of a curve L_1 on $\mathfrak{T}$ issuing from t_0 which need not be closed. If t_1 is the terminal point of the latter curve, then we can argue in the case of t_1 as we did in the case of t_0 and obtain C as the image of a definite arc L_2 of $\mathfrak{T}$ issuing from t_1 and terminating at t_2. This procedure can be continued. Since the rational nonconstant function $\varphi(t)$ takes on the same value at the points $t_0, t_1, t_2, \ldots$, the process of joining finitely many arcs $L_1, L_2, \ldots, L_n$ on $\mathfrak{T}$ yields a closed curve L whose image on $\mathfrak{R}$ is the curve C traversed n times (Figure 31).

If the genus p of $\mathfrak{R}$ were positive, then we could pick C in one of the $2p$ generating classes α of the fundamental group of $\mathfrak{R}$. But then $\alpha^n \neq \varepsilon$ for every natural number n, for this is true, by Section 4 of Chapter 2 (Volume I), even in the case of the commutative homology group. Consequently, the curve C traversed n times is not homotopic to zero on $\mathfrak{R}$. On the other hand, L is homotopic to zero on the simply connected spherical surface $\mathfrak{T}$ and has a unique continuous image on $\mathfrak{R}$. This contradiction completes the proof.

We now treat the question of obtaining all rational parametric representations of a curve of genus 0.

Theorem 3: All rational parametric representations of a curve of genus 0 are obtained by putting

$$z = \chi(t), \tag{10}$$

where z is the uniformizing parameter and $\chi(t)$ is an arbitrary nonconstant rational function.

Proof: Let $A(u, v) = 0$ be the irreducible equation of a curve of genus 0. Since the algebraic function field K associated with the curve coincides with the field of rational functions of z, we have a rational parametric representation $u = f(z)$, $v = g(z)$, and

$$z = R(u, v) \tag{11}$$

is rational in u and v. Then an arbitrary rational substitution (10) yields the rational parametric representation

$$u = f(\chi(t)) = \varphi(t), \qquad v = g(\chi(t)) = \psi(t).$$

Conversely, if we have a rational parametric representation $u = \varphi(t)$, $v = \psi(t)$ of the curve, where φ and ψ are not both constant, then (11) yields an expression for z as a rational nonconstant function of t, that is, (10). This completes the proof.

It is natural to ask when the mapping of the complex t-sphere $\mathfrak{T}$ onto the Riemann surface $\mathfrak{R}$, and so onto the complex z-sphere, effected by a rational parametric representation is one-to-one. If the rational function $\chi(t)$ is of order $n > 1$ in t, then the inversion of the substitution (10) is not unique. It follows that n must be 1 and $\chi(t)$ must be linear fractional. This agrees with the earlier result to the effect that in the case $p = 0$ the uniformizing parameter z is completely determined up to a fractional linear transformation

$$z = \frac{at + b}{ct + d} \qquad (ad - bc \neq 0).$$

Together, Theorems 2 and 3 yield the result that every rational curve $u = \varphi(t)$, $v = \psi(t)$ admits a rational parametric representation $u = f(z)$, $v = g(z)$ such that z is again rational in u and v. This is the *theorem of Lüroth*, which can also be obtained by strictly algebraic means.

The main result of the investigations of this chapter is recognition of the fact that the automorphic functions play the same role, in terms of importance and comparable properties, for algebraic curves of genus $p > 1$ as the elliptic functions for algebraic curves of genus $p = 1$. The theory of the automorphic functions is not nearly so complete and perfect as that of the elliptic functions. This is primarily because the Poincaré series have no such simple properties as the $\wp$-function. Furthermore, we have no algebraic addition theorem for the automorphic functions. We shall see in the next chapter that for

$p > 1$ the counterpart of the addition theorem for the elliptic functions is the addition theorem for the abelian functions which arise in connection with the problem of inversion of p abelian integrals of the first kind. However, these are functions of p complex variables. Also, for $p > 1$ even the inversion of a single abelian integral of the first kind yields no automorphic function.

In the case $p = 1$ introduction of the uniformizing parameter z maps a period parallelogram $\mathfrak{P}$ onto the canonically dissected Riemann surface $\mathfrak{R}$. However, if we start with a canonical dissection of $\mathfrak{R}$, then, by Section 10 of the first chapter (Volume I), we obtain as image in the z-plane a parallelogram $\mathfrak{Q}$ whose boundary is, in general, curvilinear. In this connection we pointed out (without proof) that it is possible to deform the given canonical dissection continuously in such a way that the dissection remains canonical throughout the process of deformation and the image $\mathfrak{Q}$ in the z-plane is a parallelogram $\mathfrak{P}$ bounded by straight lines.

In the case $p > 1$ the counterpart of the period parallelogram is the normal fundamental domain $\mathfrak{F}$ of a given hyperbolic polygon group Γ in the z-plane. The edges of $\mathfrak{F}$ determine a dissection of the Riemann surface $\mathfrak{R}$ which makes $\mathfrak{R}$ simply connected, but is, in general, not canonical. In particular, the number of sides of $\mathfrak{F}$ need not be $4p$. Conversely, if we start with a canonical dissection of $\mathfrak{R}$, then, under the conformal mapping of the covering surface $\mathfrak{U}$ of $\mathfrak{R}$ onto the unit disk $\mathfrak{E}$, the canonically dissected Riemann surface is carried into a portion of surface $\mathfrak{Q}$ bounded by $4p$ smooth curves which are not, in general, noneuclidean segments. The question arises, whether it is possible, in analogy to the case $p = 1$, to deform the given canonical dissection of $\mathfrak{R}$ continuously so that $\mathfrak{Q}$ becomes a noneuclidean polygon. If this happens to be the case, we shall call $\mathfrak{Q}$ a *canonical polygon*. The question just posed has not yet been answered satisfactorily in the general case.

In the case $p = 1$ we obtained all the admissible discontinuous groups by prescribing a period parallelogram whose sides yielded generators of the group. In the case $p > 1$ we started with the group itself, as in Sections 3 through 6, or with a Riemann surface, as in Section 7. In other words, we did not start with a noneuclidean polygon as fundamental region and then constructed the corresponding group Γ. It is highly significant that we can proceed in this way and, in particular, obtain all the polygon groups constructively. Geometric construction of the discontinuous groups of noneuclidean motions was investigated primarily by Poincaré and, independently of Poincaré, by Klein. In the next section we explain and clarify this approach using the example of a preassigned canonical polygon.

9. Canonical polygons

Let $\mathfrak{P}$ be a simply connected polygon with $4p$ sides, $p > 1$. We choose a definite vertex z_0, and denote the positively oriented sides by A_l, B_l, A_l', B_l',

$l = 1, \ldots, p$ [cf. Figure 43 and text of Section 2, Chapter 2 (Volume I)]. We assume that A_l and A_l' as well as B_l and B_l' have equal noneuclidean lengths and that the sum of all the $4p$ angles of $\mathfrak{P}$ is 2π.

There is exactly one noneuclidean motion α_l which carries A_l' into the side A_l traversed in the opposite direction, and exactly one noneuclidean motion β_l which carries B_l into the side B_l' traversed in the opposite direction (Figure 32). We denote the group generated by the $2p$ motions α_l, β_l by Γ, and its unit element by ε.

Theorem 1: The group Γ is hyperbolic and has $\mathfrak{P}$ as fundamental region.

Proof: The proof is in three steps.

I. For every element ρ of Γ let $\mathfrak{P}_\rho$ be the corresponding image of $\mathfrak{P}$ in $\mathfrak{E}$. We form the product $\gamma_1 \gamma_2 \cdots \gamma_p$ of commutators

$$\alpha_l \beta_l \alpha_l^{-1} \beta_l^{-1} = \gamma_l \qquad (l = 1, \ldots, p)$$

and, using the successive $4p$ factors $\alpha_1, \beta_1, \ldots, \alpha_p^{-1}, \beta_p^{-1}$ of that product, we form the $4p + 1$ elements

$$(1) \quad \rho = \varepsilon, \alpha_1, \alpha_1\beta_1, \alpha_1\beta_1\alpha_1^{-1}, \gamma_1, \gamma_1\alpha_2, \gamma_1\alpha_2\beta_2, \gamma_1\alpha_2\beta_2\alpha_2^{-1}, \gamma_1\gamma_2, \gamma_1\gamma_2\alpha_3, \ldots, \gamma_1\gamma_2, \ldots, \gamma_p.$$

If z_1, z_2, z_3, z_4 are the vertices of $\mathfrak{P}$ following z_0, then the successive mappings $\beta_1^{-1}, \alpha_1^{-1}, \beta_1, \alpha_1$ carry z_4 into z_1, z_2, z_3, z_0. This shows that for $\rho = \varepsilon, \alpha_1, \alpha_1\beta_1, \alpha_1\beta_1\alpha_1^{-1}, \gamma_1$, the corresponding five polygons $\mathfrak{P}_\rho$ share the vertex z_0, and the angles corresponding to the vertices z_0, z_3, z_2, z_1, z_4 are lined up clockwise about z_0. Also, we see that the $4p + 1$ images $\mathfrak{P}_\rho$ of $\mathfrak{P}$ under the mappings (1) fit together without gaps at z_0 to form a full neighborhood of this vertex, and that they follow each other cyclically in a clockwise sense. Since the

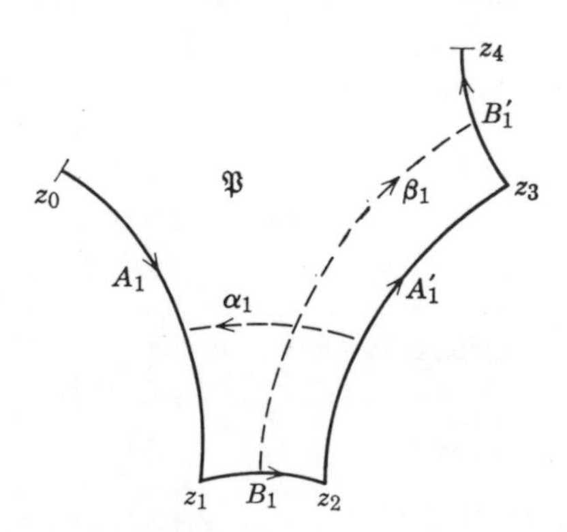

Figure 32

sum of the angles is 2π, the last polygon coincides with the initial polygon; this implies the equality

$$\gamma_1 \gamma_2 \cdots \gamma_p = \varepsilon. \tag{2}$$

By cyclic permutation of the $4p$ factors, we find in the same way that at each vertex of $\mathfrak{P}$ the $4p$ different image polygons yield a simple gapless covering of a sufficiently small neighborhood of the vertex in question. This does not prove that the polygons used in the covering do not overlap in the large.

Now we consider the polygons $\mathfrak{P}_\rho$ for arbitrary ρ in Γ and show that they cover $\mathfrak{E}$ without gaps. In fact, let $\mathfrak{M}$ be the union of all the $\mathfrak{P}_\rho$. Clearly, there exists a sufficiently small positive number d such that all the points of $\mathfrak{E}$ whose noneuclidean distance from a point of the polygon $\mathfrak{P}$ is $<d$ are in $\mathfrak{M}$. But then the same is true for all the $\mathfrak{P}_\rho$ in place of $\mathfrak{P}$. It follows that $\mathfrak{M}$ is both open and closed in $\mathfrak{E}$ and so coincides with $\mathfrak{E}$.

II. We must now show that the polygons $\mathfrak{P}_\rho$ yield a simple covering of $\mathfrak{E}$, that is, they do not overlap. To this end we employ a more general argument, which is of intrinsic interest. We consider the free group $\dot{G}$ with $2p$ independent generators $\dot{a}_l, \dot{b}_l$ $(l = 1, \ldots, p)$. Its elements are the words in these generators, their inverses $\dot{a}_l^{-1}$, $\dot{b}_l^{-1}$, and the empty word $\dot{e}$, which plays the role of the unit element. If $\dot{r}$ and $\dot{s}$ are two such words, then, by definition, $\dot{r} = \dot{s}$ if and only if it is possible to reduce the word $\dot{r}^{-1}\dot{s}$ to $\dot{e}$ by successive elimination of pairs of neighboring mutually inverse letters. It is obvious that if in place of each word $\dot{r}$ in the $\dot{a}_l, \dot{b}_l$ we take the word ρ obtained by replacing the $\dot{a}_l, \dot{b}_l$ with the Greek letters α_l, β_l, and if, in addition, we take in place of the empty word $\dot{e}$ the unit element ε, then Γ becomes a homomorphic image of $\dot{G}$. Let $\dot{\mathfrak{P}}$ be the punctured polygon obtained from $\mathfrak{P}$ by omission of all the $4p$ vertices and let us associate with each element $\dot{r}$ of $\dot{G}$ a copy $\mathfrak{P}_{\dot{r}}$ of $\dot{\mathfrak{P}}$. If $\dot{r}$ is represented in Γ by ρ, then $\mathfrak{P}_{\dot{r}}$ is to be placed over the polygon $\mathfrak{P}_\rho$ in $\mathfrak{E}$ so that corresponding sides coincide. We now make the punctured polygons $\mathfrak{P}_{\dot{r}}$ into a Riemann region $\dot{\mathfrak{B}}$ over $\mathfrak{E}$ as follows. If $\dot{r}$ and $\dot{s}$ are two distinct elements of $\dot{G}$, then the polygons $\mathfrak{P}_{\dot{r}}$ and $\mathfrak{P}_{\dot{s}}$ are to be joined along a side if and only if $\dot{r}^{-1}\dot{s}$ or $\dot{s}^{-1}\dot{r}$ is equal to one of the $2p$ generators of $\dot{G}$. In that case the corresponding element $\rho^{-1}\sigma$ or $\sigma^{-1}\rho$ is a definite generator of Γ. This determines a common side of $\mathfrak{P}_\rho$ and $\mathfrak{P}_\sigma$, and $\mathfrak{P}_{\dot{r}}$ and $\mathfrak{P}_{\dot{s}}$ are to be joined along the side lying over that common side. It is easy to see that $\dot{\mathfrak{B}}$ is connected. In fact, if, beginning at the end, we cross out one by one the letters appearing in a word $\dot{r}$ and consider the polygons on $\dot{\mathfrak{B}}$ corresponding to the resulting subwords of $\dot{r}$, then, due to the structure of $\dot{\mathfrak{B}}$, we obtain a chain of polygons, joined in pairs, which connect $\mathfrak{P}_{\dot{r}}$ with $\mathfrak{P}_{\dot{e}}$.

Let G be the group with $2p$ generators a_l, b_l $(l = 1, \ldots, p)$ and the single defining relation

$$(3) \qquad g_1 g_2 \cdots g_p = e,$$

where

$$g_l = a_l b_l a_l^{-1} b_l^{-1} \qquad (l = 1, \cdots, p),$$

and e denotes (in this proof) the unit element. If we associate with an element $\dot{r}$ in $\dot{G}$ the element r in G obtained by replacing the letters $\dot{a}_l, \dot{b}_l$ by a_l, b_l, then we obtain a homomorphic mapping of $\dot{G}$ onto G. Also, these groups are related in the manner of their isomorphic replicas $\dot{\Gamma}$ and Γ which appeared in Section 3 of Chapter 2 (Volume I) as the fundamental groups of $\dot{\mathfrak{R}}_p$ and $\mathfrak{R}_p$. Two polygons $\mathfrak{P}_{\dot{r}}$ and $\mathfrak{P}_{\dot{t}}$ are to be identified if and only if the elements r and t corresponding to $\dot{r}$ and $\dot{t}$ in $\dot{G}$ are equal. Since, by (2), the concrete group Γ is also a homomorphic image of G, the equality $r = t$ surely implies the equality $\rho = \tau$. This means that, before the identification, the polygons $\mathfrak{P}_{\dot{r}}$ and $\mathfrak{P}_{\dot{t}}$ must have both lain over $\mathfrak{P}_\rho$. We denote the punctured polygon over $\mathfrak{P}_\rho$ resulting from the identification by $\dot{\mathfrak{P}}_r$. It must be shown that this joining of complete polygons again yields a Riemann region $\mathfrak{B}_0$ over $\mathfrak{E}$. In view of our construction, two polygons $\dot{\mathfrak{P}}_r$ and $\dot{\mathfrak{P}}_s$ have a side in common if and only if there are two elements $\dot{t}$, $\dot{u}$ with $r = t$, $s = u$, for which $\mathfrak{P}_{\dot{t}}$ and $\mathfrak{P}_{\dot{u}}$ have a corresponding side in common. But this means that $r^{-1}s$ or $s^{-1}r$ must be a generator of G. It follows that in the process of identification of the $\mathfrak{P}_{\dot{r}}$ and $\mathfrak{P}_{\dot{s}}$ complete neighborhoods are joined at corresponding boundary points as well. Since $\dot{\mathfrak{B}}$ is connected, $\mathfrak{B}_0$ is surely connected. Hence $\mathfrak{B}_0$ is indeed a Riemann region constructed out of the individual polygons $\dot{\mathfrak{P}}_r$.

III. The neighboring polygons $\dot{\mathfrak{P}}_r$ with

$$(4) \quad r = e,\, a_1,\, a_1 b_1,\, a_1 b_1 a_1^{-1},\, g_1,\, g_1 a_2,\, g_1 a_2 b_2,\, g_1 a_2 b_2 a_2^{-1},$$
$$g_1 g_2,\, g_1 g_2 a_3, \ldots, g_1 g_2 \cdots g_p$$

lie about the vertex of $\dot{\mathfrak{P}}_e$ over z_0 over the polygons $\mathfrak{P}_\rho$ which correspond to them by (1). In both cases the polygons are arranged in the same cyclic order. By (3), the last of the polygons $\dot{\mathfrak{P}}_r$ coincides with the first. It follows that if the vertices of the polygons $\dot{\mathfrak{P}}_r$ lying over z_0 are included as a common point, then this point has a full schlicht neighborhood on $\mathfrak{B}_0$. By means of the correspondence from x to sx, with arbitrary fixed s and variable x in G, this argument carries over to the vertices of all the polygons $\dot{\mathfrak{P}}_s$. Let $\mathfrak{P}_s$ denote the complete polygon obtained by inclusion of the $4p$ vertices of $\dot{\mathfrak{P}}_s$. Then all the $\mathfrak{P}_s$ form a Riemann region $\mathfrak{B}$ over $\mathfrak{E}$ which goes over into

$\mathfrak{B}_0$ as a result of puncturing all the vertices of the polygons. It is clear that $\mathfrak{B}$ has no branch points. Furthermore, there exists a sufficiently small positive number d such that at every point of $\mathfrak{B}$ and in every direction it is possible to lay on $\mathfrak{B}$ a unique noneuclidean segment of length d. Now let C be a curve in $\mathfrak{E}$ issuing from z_0 which is made up of finitely many noneuclidean segments. Then there is exactly one polygonal curve L on $\mathfrak{B}$ which lies over C and issues from the vertex of $\mathfrak{P}_e$ over z_0. In the sequel we identify $\mathfrak{P}_e$ with $\mathfrak{P}$.

At the end of Section 8, Chapter 2 (Volume I), we showed that there exists a function element which is regular in the vicinity of z_0 and which yields by the process of analytic continuation a function $f(z)$ with the prescribed Riemann surface $\mathfrak{B}$. Since C was an arbitrary polygonal curve with initial point z_0, the given function element at z_0 in $\mathfrak{E}$ can be continued analytically along C; this follows because L lies completely on $\mathfrak{B}$. By the monodromy theorem, the function $f(z)$ must already be single-valued on $\mathfrak{E}$, so that $\mathfrak{B}$ coincides with $\mathfrak{E}$. This fact can also be proved directly without the introduction of the function $f(z)$ by making use of the reasoning employed in the proof of the monodromy theorem to show quite generally that every unramified unbounded covering surface of a simply-connected schlicht region of the z-plane coincides with that region.

Let ρ and σ in Γ correspond to two arbitrary elements r and s in G. If $\rho = \sigma$, then the polygons $\mathfrak{P}_\rho$ and $\mathfrak{P}_\sigma$ are the same. But then both $\mathfrak{P}_r$ and $\mathfrak{P}_s$ lie over $\mathfrak{P}_\rho$. Since $\mathfrak{B}$ is schlicht, $\mathfrak{P}_r$ and $\mathfrak{P}_s$ are the same and we have $r = s$. It follows that Γ is a faithful representation of the abstract group G with $2p$ generators and the single defining relation (3). On the other hand, $\mathfrak{P}$ is now seen to be a fundamental region of Γ and the images $\mathfrak{P}_\rho$ of $\mathfrak{P}$ yield a simple covering of $\mathfrak{E}$. Since $\mathfrak{P}$ is compact in $\mathfrak{E}$, Γ is a polygon group without parabolic elements. If Γ had a fixed point in $\mathfrak{E}$, then $\mathfrak{P}$ would contain a fixed point of Γ, which would have to be a vertex. For the $4p$ different mappings ρ in (1) the polygons $\mathfrak{P}_\rho$ yield a full neighborhood of z_0, and the mappings ρ^{-1} carry z_0 into all the $4p$ vertices of $\mathfrak{P}$. This rules out the existence of fixed points and shows that Γ is hyperbolic. Now the proof is complete.

It should be pointed out that, in spite of the fact that the fundamental region $\mathfrak{P}$ of Γ is a noneuclidean polygon it need not be a normal polygon. If $\mathfrak{P}$ is a regular noneuclidean polygon, that is, if it has equal angles and sides of the same noneuclidean length, then the perpendicular bisectors of the sides intersect in a point ζ, and then $\mathfrak{P}$ is a normal fundamental region of Γ with center ζ. Since the sum of the angles must be 2π, every angle of $\mathfrak{P}$ must, in the present case, be equal to $\pi/(2p)$. To show the existence of such a regular noneuclidean polygon, we consider a regular noneuclidean polygon with $4p$ vertices with center at $z = 0$ and vertex at $z = r$ on the positive real axis. If r increases continuously from 0 to 1, then the sum of the angles of the polygon decreases continuously and monotonically from $(4p - 2)\pi$ to 0. It follows that there exists a positive $r < 1$ for which the sum of the angles is exactly 2π. By Theorem 4, Section 2, the noneuclidean area of this polygon $\mathfrak{P}$ is

$$I(\mathfrak{P}) = (4p - 2)\pi - 2\pi = 4(p - 1)\pi. \tag{5}$$

It is not difficult to compute explicitly the value r and to determine the vertices of $\mathfrak{P}$ as well as the corresponding noneuclidean motions α_l, β_l $(l = 1, \ldots, p)$. This gives the group Γ for the case under consideration.

The preceding results answer a question left open in Section 3 of the second chapter (Volume I). There we were concerned with the construction of a plane region $\mathfrak{B}$ which is the topological image of the covering surface of a compact Riemann region $\mathfrak{R}$ of positive genus p, with $\mathfrak{R}$ itself represented, after canonical dissection, by a regular euclidean polygon $\mathfrak{N}$ of $4p$ sides. In the case $p = 1$, $\mathfrak{B}$ was readily obtained as the covering of the plane with a net of squares which are euclidean-congruent to $\mathfrak{N}$. In the case $p > 1$ the euclidean polygons used to construct $\mathfrak{B}$ were obtained from $\mathfrak{N}$ by means of none too transparent distortions. Using noneuclidean geometry, however, we can proceed in analogy to the case $p = 1$ by taking $\mathfrak{N} = \mathfrak{P}$ and $\mathfrak{B} = \mathfrak{E}$.

If we start with an arbitrary noneuclidean polygon $\mathfrak{P}$ with $4p$ sides and angles of the prescribed kind, then, after we choose a fixed vertex z_0, the $2p$ motions α_l, β_l $(l = 1, \ldots, p)$ generate a hyperbolic group Γ with $\mathfrak{P}$ as fundamental region. The automorphic functions associated with Γ form a field K. After we choose an arbitrary nonconstant function $w = f(z)$ in K, K can be described as the field of algebraic functions of w which are meromorphic on the compact Riemann surface $\mathfrak{R}$ which arises over the w-sphere as a result of the mapping of $\mathfrak{P}$. Furthermore, $\mathfrak{E}$ is mapped conformally onto the covering surface $\mathfrak{U}$ of $\mathfrak{R}$, the sides of $\mathfrak{P}$ determine on $\mathfrak{R}$ a fixed canonical dissection, and Γ is a faithful representation of the fundamental group of $\mathfrak{R}$ which is isomorphic to G. It follows that K has genus p, $\mathfrak{P}$ is a canonical polygon, and (5) coincides with (6) of Section 7.

In deriving Theorem 1 we had to prove that Γ is the group with $2p$ generators α_l, β_l $(l = 1, \ldots, p)$ and the single defining relation (2). In the proof we used, among others, the monodromy principle. Of the existing texts and papers on the theory of automorphic functions which deal with the question under consideration none contains a satisfactory proof, and some contain crude errors.

Under the stated assumptions the canonical polygon $\mathfrak{P}$ can be chosen arbitrarily in $\mathfrak{E}$, and then the field K is uniquely determined. This suggests the natural question as to the number of independent parameters which enter into the choice of $\mathfrak{P}$, and so of K, when the genus $p > 1$ is prescribed.

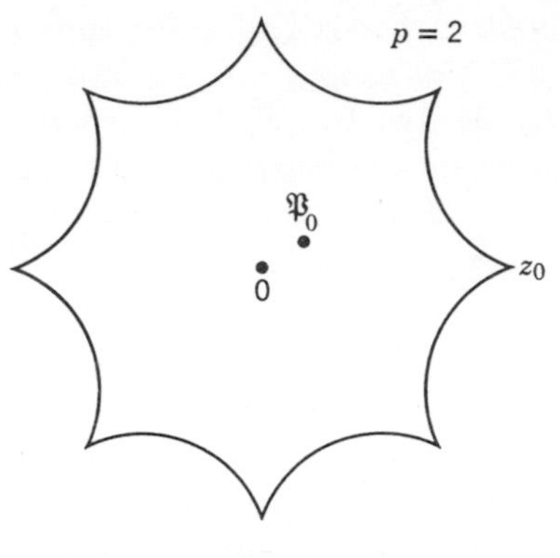

Figure 33

The general treatment of this question leads to difficult and partly unsolved problems. We cannot go deeply into these matters and will, instead, be satisfied with a few comments.

We now consider only those canonical polygons which lie sufficiently close to a fixed starting polygon $\mathfrak{P}_0$ with vertex z_0. One could, for example take $\mathfrak{P}_0$ to be the regular noneuclidean polygon with $4p$ sides and angle sum 2π whose center is at the origin and whose vertex z_0 lies on the positive real axis (Figure 33). We take for α_l, β_l $(l = 1, \ldots, p)$ noneuclidean motions which lie close to those determined by $\mathfrak{P}_0$ and likewise satisfy the condition (2). Suppose a first vertex lying close to z_0 chosen in an arbitrary manner. Then the $2p$ generators determine a unique polygon $\mathfrak{P}$ whose angle sum is also 2π. If we are already presented with the conformal mapping $w = f(z)$ of $\mathfrak{E}$ onto $\mathfrak{U}$, then a shift in the vertex z_0 merely signifies a shift in the point on the Riemann surface which serves as the origin of the canonical dissection. It therefore suffices to keep the point z_0 fixed. If instead of $\mathfrak{E}$ we work with the upper half plane, then every noneuclidean motion is given by

$$(6) \qquad v = \frac{az + b}{cz + d} \qquad (ad - bc = 1)$$

with real a, b, c, d. Since in the hyperbolic case $(a + d)^2 > 4$, we can, for the purpose of unique determination of the motion, consider as independent parameters locally either a, b, c or b, c, d. Accordingly, the $2p$ generators α_l, β_l, $(l = 1, \ldots, p)$ of Γ contain $6p$ real parameters, which, by (2), satisfy three equations. While it is not difficult to show these equations independent, we shall pass over this point. We note, furthermore, that for every fixed noneuclidean motion μ, the group $\mu^{-1}\Gamma\mu$, conjugate to Γ in Ω, yields the same field K as Γ. This is so because conjugation of Γ is tantamount to a different choice of the uniformizing parameter z. In particular, it is possible to choose μ so that three of the six real parameters in $\mu^{-1}\alpha_1\mu$ and $\mu^{-1}\beta_1\mu$ take on prescribed values. For example, referring to (6), we could prescribe the values $b = 0$, $c = 0$ for the first motion and the value $c = 1$ for the second and call the generators *normalized.* Now the count yields $6p - 6$ real parameters on which the field K depends. What remains to be considered is whether, conversely, these parameters are uniquely determined, at least locally, by K. If α_l^*, β_l^* $(l = 1, \ldots, p)$ are also normalized generators of a group Γ^*, and $\mathfrak{P}^*$ is the corresponding canonical polygon, then the corresponding field coincides with K if and only if $\Gamma^* = \mu^{-1}\Gamma\mu$ for an appropriate μ in Ω. If $\mu = \varepsilon$ or, more generally, if μ is an element of Γ, then $\mathfrak{P}^*$ and $\mathfrak{P}$ coincide in a sufficiently small neighborhood of $\mathfrak{P}_0$, and, in view of the discreteness of Γ, the parameters are uniquely determined. The essential difficulty lies in the fact that μ need not lie in Γ. That is why the attempt to introduce $6p - 6$ real or $3p - 3$ complex (so-called) moduli of an algebraic function field of genus $p > 1$ in the indicated manner is bound to fail.

On the other hand, in the case $p = 1$ the corresponding question admits of a straightforward discussion. As was shown in the preceding section, the field K determines the lattice of periods ω to within an arbitrary complex factor α^{-1}. This enables us to choose one of the two fundamental periods ω_1, ω_2, say ω_1, equal to 1. Since the fundamental periods are determined by the period lattice only up to a unimodular substitution, then, with the normalization $\omega_1 = 1$, $\omega_2 = \tau$,

$$\omega_2 = \frac{a\tau + b}{c\tau + d} \tag{7}$$

is also admissible; here a, b, c, d are integers and $ad - bc = \pm 1$. The ambiguity of sign can be resolved in favor of the plus sign if we suppose the second fundamental period to lie, like τ, in the upper half plane $\mathfrak{H}$. If we call two points in the upper τ-half plane *equivalent* whenever they are connected by a modular substitution (7), then this establishes a one-to-one correspondence between the classes of equivalent τ and the algebraic function fields of genus 1. Every such class is then to be viewed as the *modulus of the corresponding field*, which is why the group defined by (7) with integer a, b, c, d and $ad - bd = 1$ is called the *modular group* or, more precisely, the *elliptic modular group*. If we wish to have a particular representative of a class as modulus, then we can choose for this purpose the reduced value of $\tau = \xi + i\eta$ defined, according to Theorem 3 in Section 9, Chapter 1 (Volume I) and the subsequent improvement, by the inequalities $\xi^2 + \eta^2 \geqslant 1$, $-\frac{1}{2} \leqslant \xi < \frac{1}{2}$ and the side condition $\xi \leqslant 0$ for $\xi^2 + \eta^2 = 1$. This choice yields the familiar fundamental region of the modular group. The quantity j defined in (6), Section 8, depends, by (3), Section 8, solely on the period lattice and is homogeneous of dimension 0 in ω_1, ω_2. It follows that $j = j(\tau)$, considered on $\mathfrak{H}$, is a regular function of τ alone which has the invariance property

$$j\left(\frac{a\tau + b}{c\tau + d}\right) = j(\tau)$$

under all modular substitutions, and is therefore called the *elliptic modular function.* Its importance consists in the fact that, in view of Theorem 4, Section 8, its values $j(\tau)$ separate the various classes of equivalent τ.

Our earlier remark in connection with the problem of moduli for $p > 1$ suggests the investigation of the equation $\Gamma = \mu^{-1}\Gamma\mu$, where Γ is a hyperbolic polygon group and μ is a noneuclidean motion. Every solution μ yields an automorphism and those automorphisms form a subgroup Δ of Ω. In particular, if μ commutes with all the elements of Γ, then, by Theorem 4, Section 1, we must have $\mu = \varepsilon$; for, otherwise, contrary to the noncommutative character of Γ, μ would have the same fixed points as all the hyperbolic elements of Γ. It follows that different solutions μ determine different automorphisms of Γ. It is clear that Γ is a normal subgroup of Δ; in fact, Δ is the largest subgroup of Ω containing Γ as a normal subgroup.

Theorem 2: Let Δ be a subgroup of Ω which contains the hyperbolic polygon group Γ as normal subgroup, and let p be the genus of the field of automorphic functions associated with Γ. Then Δ is discontinuous and the index of Γ in Δ is not greater than $84(p - 1)$.

Proof: If the group Δ were not discontinuous in $\mathfrak{E}$, then, by Theorem 2, Section 3, it would not be a discrete subgroup of Ω. By Theorem 1 of the same section, there would exist elements $\mu \neq \varepsilon$ in Δ which lie arbitrarily close to the unit element ε. Since Γ is discontinuous and therefore discrete, it would follow that, for a given α in Γ and μ in Δ lying sufficiently close to ε, the commutator $\mu^{-1}\alpha\mu\alpha^{-1}$ in Γ coincides with ε. But then there would exist a $\mu \neq \varepsilon$ in Δ which, after choice of $2p$ generators of Γ, would commute with the latter, and so with all the elements of Γ. As noted above, this is a contradiction. It follows that Δ is discontinuous.

Continuing, let $\mathfrak{F}$ be a compact fundamental region of Γ and ξ a point of $\mathfrak{E}$. In every coset of Δ relative to Γ it is possible to choose an element μ such that the mapping μ carries ξ into a point of $\mathfrak{F}$. Since Δ is discontinuous, the number of such μ must be finite. This number, however, is the index j of Γ in Δ. If $\mathfrak{G}$ is a fundamental region of Δ, then the union of the j particular images $\mathfrak{G}_\mu$ of $\mathfrak{G}$ is likewise a fundamental region $\mathfrak{F}^*$ of Γ. By Theorem 5, Section 4,

$$I(\Gamma) = I(\mathfrak{F}) = I(\mathfrak{F}^*) = jI(\mathfrak{G});$$

by (6), Section 7,

$$I(\mathfrak{F}) = 4(p - 1)\pi,$$

and by Theorem 2, Section 4,

$$I(\mathfrak{G}) \geqslant \frac{\pi}{21}.$$

This implies

$$j = \frac{I(\mathfrak{F})}{I(\mathfrak{G})} \leqslant 84(p-1),$$

which proves the remaining part of the theorem. We note that in proving the second part of the theorem we made no use of the fact that Γ is normal in Δ.

Theorem 3: For $p > 1$ the number of conformal mappings which carry a given compact Riemann region of genus p into itself is invariably finite and, in fact, not greater than $84(p-1)$.

Proof: Let $\mathfrak{R}$ be the given compact region over the w-sphere and let $w = f(z)$ be a conformal mapping of the unit disk $\mathfrak{E}$ onto the covering surface $\mathfrak{U}$ of $\mathfrak{R}$. Furthermore, let Γ be the hyperbolic polygon group which represents the fundamental group of $\mathfrak{R}$. For every element α of Γ, $s = z_\alpha$ denotes a noneuclidean motion. If, once more, $\mu^{-1}\Gamma\mu = \Gamma$ for an element μ in Ω, then, like α, $\mu\alpha\mu^{-1} = \beta$ is in Γ and

$$s_\mu = z_{\mu\alpha} = z_{\beta\mu} = (z_\mu)_\beta,$$

so that both z_μ and s_μ are equivalent with respect to Γ. It follows that the mapping $s = z_\alpha$ not only effects a conformal mapping of $\mathfrak{U}$ onto itself but also a conformal mapping of $\mathfrak{R}$ onto itself. Conversely, every conformal mapping of $\mathfrak{R}$ onto itself yields a conformal mapping of $\mathfrak{U}$ onto itself determined up to a covering transformation, and thus a noneuclidean motion μ which commutes with Γ and is determined up to a factor in Γ. The solutions μ of $\mu^{-1}\Gamma\mu = \Gamma$ form a subgroup Δ of Ω containing Γ as a normal subgroup. It follows that the conformal mappings of $\mathfrak{R}$ onto itself form a group isomorphic to the factor group Δ/Γ. By Theorem 2 this factor group is finite and its order is no greater than $84(p-1)$.

Let $g = g(w)$ be an algebraic function with Riemann surface $\mathfrak{R}$ and $P(g, w) = 0$ the corresponding irreducible algebraic equation. Then it is clear that the conformal mappings of $\mathfrak{R}$ onto itself are just the birational transformations which carry the algebraic curve $P(g, w) = 0$ into itself. It follows that if the genus p of the curve is greater than 1, then, by Theorem 3, the number of these birational transformations is always finite, and, in fact, not greater than $84(p-1)$. The finiteness was established by Poincaré, who relied on preliminary work of Schwarz and Klein. The specific upper bound was subsequently found by Hurwitz. Without going into the matter we wish to point out that, as shown by Hurwitz, the upper bound $84(p - 1)$ is actually achieved for $p = 3$.

It remains to discuss birational transformations for the hitherto excluded values $p = 0$ and $p = 1$. In this connection we note that the study of the connection between the conformal mappings of $\mathfrak{R}$ onto itself and the automorphisms of the group Γ carried out under the assumption $p > 1$ is directly applicable to the cases $p = 0$ and $p = 1$. In the case $p = 0$, $\mathfrak{E}$ is replaced by the z-sphere, Γ is the identity, and Δ is the group of all fractional linear transformations with arbitrary complex coefficients. It follows that this group contains three independent complex parameters. A corresponding assertion holds for the birational mappings which carry an algebraic curve of genus 0 into itself, so that it is possible to find among them a mapping which carries three distinct given points on the curve into three other distinct and arbitrarily assigned points on the curve. In the case $p = 1$, $\mathfrak{E}$ is replaced by the z-plane, and Γ consists of the translations $s = z + \omega$ with two generators ω_1, ω_2, where $\omega_2\omega_1^{-1} = \tau$ lies in the upper half plane. If μ is a conformal mapping of the z-plane onto itself, that is, a linear transformation $s = mz + r (m \neq 0)$ with constant m and r, then $\mu\Gamma\mu^{-1}$ consists of the translations $s = z + m\omega$. This group of translations coincides with the original group of translations if and only if the two equations

$$m\omega_2 = a\omega_2 + b\omega_1, \qquad m\omega_1 = c\omega_2 + d\omega_1 \qquad (ad - bc = \pm 1)$$

hold for integers a, b, c, d. If that is the case, then we have for the ratio τ of the periods

$$m\tau = a\tau + b, \qquad m = c\tau + d, \qquad \tau = \frac{a\tau + b}{c\tau + d},$$

so that, in particular, $ad - bc = 1$. This implies for τ the equation

$$c\tau^2 + (d - a)\tau - b = 0. \tag{8}$$

Now we distinguish two cases. If (8) holds identically in τ, that is, if $b = c = 0$ and $a = d = \pm 1$, then $m = \pm 1$, and the linear transformation $s = mz + r$ reduces to $s = z + r$ or $s = -z + r$. For every complex r, these yield birational transformations which carry the given algebraic curve of genus 1 into itself. If we start with that curve in the normal form

$$h^2 = 4f^3 - g_2 f - g_3 \tag{9}$$

and put $f = \wp(z)$, $h = \wp'(z)$, then there correspond to the transition from z to $z + r$ the formulas of the addition theorem of the $\wp$-function which, by Section 15 of the first chapter (Volume I), express the values $\wp(z + r)$ and $\wp'(z + r)$ rationally in terms of $\wp(z)$, $\wp'(z)$, $\wp(r)$ and $\wp'(r)$. If we note that $\wp(-z) = \wp(z)$ and $\wp'(z) = -\wp'(z)$, then we obtain analogous formulas for the transition from z to $-z + r$. The translations $s = z + r$ form a group which contains the arbitrary complex parameter r.

It remains to consider the case when (8) does not hold identically in τ. Then (8) represents a quadratic equation with rational integer coefficient for the determination of the number τ in the upper half plane. Then τ is an imaginary quadratic surd and $c \neq 0$, so that τ and $m = c\tau + d$ belong to the same number field. Since m must satisfy the quadratic equation

$$m^2 - (a + d)m + 1 = 0$$

and cannot be real, it follows that $(a + d)^2 < 4$ and m is a fourth or sixth root of unity. In addition, we can assume that the number $\tau = \xi + i\eta$ is reduced in the sense of Theorem 3 in Section 9 of Chapter 1 (Volume I) and the subsequent remark, that is, that it belongs to the fundamental region of the modular group defined by the inequalities $\xi^2 + \eta^2 \geqslant 1$, $-\frac{1}{2} \leqslant \xi < \frac{1}{2}$, with the side condition $\xi < 0$ for $\xi^2 + \eta^2 = 1$. If $m = \pm i = c\tau + d$ with integer c, d, then we have $c\xi + d = 0$, $c\eta = \pm 1$, that is, $c = \pm 1$, $d = 0$ and $\tau = i$. If $m = \pm e^{2\pi i/3}$ or $m = \pm e^{\pi i/3}$, then we have $c\xi + d = \pm\frac{1}{2}$, $c\eta = \pm\sqrt{3}/2$, that is, $c = \pm 1$, $d = 0$ or 1, and $\tau = e^{2\pi i/3}$. In the case $\tau = i$ we have

$$g_3 = \sum_{\omega \neq 0} \omega^{-6} = \sum_{\omega \neq 0} (m\omega)^{-6} = m^{-6} g_3 = -g_3 = 0, \qquad j(\tau) = 1,$$

and in the case $\tau = e^{2\pi i/3}$ we have

$$g_2 = \sum_{\omega \neq 0} \omega^{-4} = \sum_{\omega \neq 0} (m\omega)^{-4} = m^{-4} g_2 = e^{\mp 2\pi i/3} g_2 = 0, \qquad j(\tau) = 0.$$

In both cases we obtain by means of the linear mapping $s = mz$ the birational transformation

$$\wp(mz) = m^{-2}\wp(z), \qquad \wp'(mz) = m^{-3}\wp'(z),$$

which leaves (9) invariant and can be combined with the mapping $s = z + r$ already treated.

All in all, the birational transformations of an algebraic curve of genus 1 into itself are obtained from the transformations $s = mz + r$ of the uniformizing parameter for arbitrary complex r, where m is a sixth root of unity in the case $j = 0$, a fourth root of unity in the case $j = 1$, and a second root of unity otherwise.

4

Abelian integrals

1. Reduction

Let K be an algebraic function field of one independent complex variable x. K can be generated by the adjunction of a function y to the field of rational functions of x, where x and y are connected by means of an irreducible equation $P(x, y) = 0$ in which y actually appears. Differentiating with respect to x we obtain

$$P_x + P_y \frac{dy}{dx} = 0$$

and see that the derivative of y with respect to x is in K. More generally, if $\varphi = Q(x, y)$ is an arbitrary element of K, that is, a rational function of x and y, then

$$\frac{d\varphi}{dx} = Q_x + Q_y \frac{dy}{dx},$$

and so the derivative of φ with respect to x is in K.

If φ and ψ are two functions in the field K, then we shall call the expression $\varphi\, d\psi$ an *abelian differential.* In this definition x is not distinguished as an independent variable. If we want to distinguish x, we must bear in mind that

$$d\psi = \frac{d\psi}{dx}\, dx,$$

and, putting, for brevity,

$$\varphi \frac{d\psi}{dx} = R(x, y),$$

we obtain the equality

$$\varphi\, d\psi = R(x, y)\, dx,$$

where $R(x, y)$ is in K. Conversely, it is clear that if R is an element of K, then $R\, dx$ is, by definition, an abelian differential. This shows that abelian differentials could have been defined directly as expressions of the form $R\, dx$. Our original definition is nevertheless preferable, since it does not depend on the choice of the independent variable and is, therefore, invariant under birational transformations.

Let $\mathfrak{R}$ be the Riemann surface of y as a function of x. Let ζ be the local parameter at a point $\mathfrak{x}$ of $\mathfrak{R}$. The functions $x = x(\zeta)$ and $y = y(\zeta)$ and, more generally, $R(x, y)$, are meromorphic at $\zeta = 0$; the same is therefore true of $dx/d\zeta$ and of the function

$$R(x, y)\frac{dx}{d\zeta} = \chi(\zeta).$$

Thus we have for the abelian differential

$$R(x, y)\,dx = \chi(\zeta)\,d\zeta, \tag{1}$$

where $\chi(\zeta)$ is either regular at $\zeta = 0$ or has a pole there. In the first case we say that the abelian differential is regular at the point $\mathfrak{x}$ of $\mathfrak{R}$. If $\chi(\zeta)$ vanishes to order l at $\zeta = 0$, then $\mathfrak{x}$ is called a *zero of order* l of the abelian differential. There is a similar definition of *pole* and *order of pole*. In the case of a pole of order l we have in a neighborhood of $\mathfrak{x}$ the following power series expansion:

$$\chi(\zeta) = \sum_{n=-l}^{\infty} c_n\zeta^n \qquad (c_{-l} \neq 0), \tag{2}$$

whose coefficient c_{-1} is called the *residue of the differential* $R(x, y)\,dx$ *at the point* $\mathfrak{x}$. We shall call the expression

$$h(\zeta)\,d\zeta = \left(\sum_{n=-l}^{-1} c_n\zeta^n\right) d\zeta \tag{3}$$

the *principal part of the abelian differential at the point* $\mathfrak{x}$.

If we introduce by means of a birational transformation new variables u and v, with u the independent variable, then this induces a conformal mapping of the Riemann surface $\mathfrak{R}$ onto the Riemann surface $\mathfrak{S}$ of v as a function of u. If ω is the local parameter on $\mathfrak{S}$ at the image point $\mathfrak{u}$ of $\mathfrak{x}$, then ω is a regular function of ζ in the vicinity of $\zeta = 0$ with nonvanishing derivative at $\zeta = 0$; so that

$$\omega = a_1\zeta + a_2\zeta^2 + \cdots, \qquad a_1 \neq 0; \qquad \omega^{-k} = (a_1\zeta)^{-k} + \cdots \qquad (k = 1, 2, \ldots).$$

It follows that the order of a zero or pole of an abelian differential depends only on K, and not on the choice of x and y. Furthermore,

$$\omega^{-k-1}\,d\omega = \frac{-1}{k}\,d(\omega^{-k}) = \frac{-1}{k}\,d(a_1^{-k}\zeta^{-k} + \cdots) \qquad (k = 1, 2, \ldots),$$

so that this expression has residue 0 with respect to ζ, because no power of ζ yields a ζ^{-1} term upon differentiation. On the other hand,

$$\omega^{-1}\,d\omega = (a_1\zeta + \cdots)^{-1}(a_1\,d\zeta + \cdots) = \zeta^{-1}\,d\zeta + \cdots$$

has residue 1. It follows that the residue of an abelian differential at a pole does not depend on the choice of the independent variable x. It is easy to see that for $l > 1$ this invariance does not carry over to the remaining coefficients $c_{-2}, \ldots, c_{-l}$ of the principal part (3).

Theorem 1: The sum of the residues of an abelian differential is always zero.

Proof: We suppose the Riemann surface $\mathfrak{R}$ canonically dissected and mapped conformally by means of the uniformizing variable s onto a schlicht domain $\mathfrak{B}$. Let $\mathfrak{R}$ have genus p. If $p > 0$, the positively traversed boundary C of $\mathfrak{B}$ consists of $2p$ pairs of sides which are identified on $\mathfrak{R}$, and have opposite orientation on $\mathfrak{R}$. We may suppose that the canonical dissection does not pass through any pole of the given abelian differential $R\,dx$. By the residue theorem, the sum of the residues has the value

$$\sigma = \frac{1}{2\pi i}\int_C R(x, y)\frac{dx}{ds}\,ds. \tag{4}$$

Since, apart from sign, the differential $R(x, y)\,dx$ has the same value at corresponding points of paired sides, the contribution of each pair of sides to the integral is 0, and, consequently, $\sigma = 0$. In the case $p = 0$, $\mathfrak{B}$ is the s-sphere, and we can normalize s so that $s = \infty$ is not a pole of the abelian differential $R\,dx$. Then we take C to be the circle $|s| = \rho$ with sufficiently large radius ρ. Letting ρ tend to ∞ and passing to the limit in (4) we again obtain $\sigma = 0$. This completes the proof.

When there is no danger of confusion we shall use the shorter term differential in place of abelian differential. A differential is said to be *of the first kind* if it is regular on all of $\mathfrak{R}$, that is, if it has no poles. A differential is said to be *of the second kind* if it has at least one pole and if, in addition, its residue at each pole is 0. Finally, a differential is said to be *of the third kind* if it has at least one nonzero residue. In the last case the differential must, by Theorem 1, have at least two poles with nonzero residues. The preceding discussion makes it clear that our classification of differentials depends only on K and is independent of the choice of the independent variable x. It is clear that if $R_1(x, y)\,dx$ and $R_2(x, y)\,dx$ are differentials, then so is $(\lambda_1 R_1 + \lambda_2 R_2)\,dx$ for every choice of complex numbers λ_1 and λ_2. This motivates the problem of obtaining an additive decomposition of the differentials, their so-called *reduction*, into certain normal forms. Before we treat this problem we must define the notion of an abelian integral and its periods.

Let $\mathfrak{x}_0$ be a given point on the Riemann surface $\mathfrak{R}$ at which the differential $R(x, y)\,dx$ is regular; let $\mathfrak{x}$ be another point on the Riemann surface $\mathfrak{R}$; and let x_0 and x be the projections of $\mathfrak{x}_0$ and $\mathfrak{x}$ on the x-sphere. If we employ rectilinear integration from $\mathfrak{x}_0$ to $\mathfrak{x}$ on $\mathfrak{R}$, then the function

$$w = w(x) = \int_{x_0}^{x} R(x, y)\, dx \tag{5}$$

is regular in a sufficiently small neighborhood of $\mathfrak{x}_0$ on $\mathfrak{R}$. If C is a curve on $\mathfrak{R}$ issuing from $\mathfrak{x}_0$ and passing through no poles of the integrand, the $w(x)$ can be continued analytically along C and has the value

$$w(C) = \int_C R(x, y)\, dx$$

at the terminal point $\mathfrak{x}$ of C. The function element $w(x)$ given by (5) and, more generally, the value $w(C)$, is called an *abelian integral.* Here we point out that $w(x)$ and $w(C)$ are defined by means of definite integrals, so that $w(x_0) = 0$. If we change the initial point $\mathfrak{x}_0$, then we must add a suitable constant to w. The *type of an abelian integral*, or, more briefly, *of an integral*, is defined to be the same as the type of the corresponding abelian differential $R(x, y)\, dx$. If the curve C is closed, then $w(C)$ is called a *period* of the integral w. If C_1 and C_2 are two curves on $\mathfrak{R}$ from $\mathfrak{x}_0$ to $\mathfrak{x}$, then the difference

$$w(C_1) - w(C_2) = w(C_1 C_2^{-1})$$

is a period. It follows that any two function elements w defined by analytic continuation of $w(x)$ to a point $\mathfrak{x}$ along arbitrary paths differ only by an additive period, and we have

$$dw = R(x, y)\, dx, \qquad w(C) = \int_C dw, \qquad w(x) = \int_{x_0}^{x} dw.$$

If ζ is a local parameter at a pole $\mathfrak{p}$ of the abelian differential dw, then we obtain from (2) as indefinite integral

$$\int dw = \int \chi(\zeta)\, d\zeta = \sum_{n=-l}^{-2} \frac{c_n}{n+1} \zeta^{n+1} + c_{-1} \log \zeta + c_0 \zeta + \cdots \tag{6}$$

$$(c_{-l} \neq 0);$$

and so, by (1), the behavior of the abelian integral w for $\mathfrak{x} \to \mathfrak{p}$ is determined by the behavior of the series in (6). We see that if the residue $c_{-1} = 0$, then $\mathfrak{p}$ is a pole of order $l - 1$. If $c_{-1} \neq 0$, then the expansion contains the term $c_{-1} \log \zeta$, and $\mathfrak{p}$ is a logarithmic branch point, which must not be included in the Riemann surface of w. It follows that the integrals of the first kind are finite everywhere regardless of the choice of the path of integration, and that the poles of the integrals of the second kind are precisely the poles of their integrands. In the case of integrals of the third kind, there must necessarily appear logarithmic singularities at all poles of the integrand which have nonvanishing residues.

The integral w is single-valued in the large on $\mathfrak{R}$ if and only if all the periods are 0. In that case, in particular, all the residues are 0, so that the integral is of the first or second kind, and, consequently, a meromorphic function φ on $\mathfrak{R}$. Now, such a function belongs to the algebraic function field K and vanishes at the point $\mathfrak{x}_0$. If w is of the first kind, then φ is regular on all of $\mathfrak{R}$, and, therefore, $\varphi = 0$. This means that there are no nontrivial single-valued integrals of the first kind on $\mathfrak{R}$, and the single-valued integrals of the second kind on $\mathfrak{R}$ are nonconstant functions in K which vanish at $\mathfrak{x}_0$.

To investigate the periods of an integral of first or second kind we introduce a canonical dissection of $\mathfrak{R}$ involving p pairs of crosscuts A_k, B_k ($k = 1, \ldots, p$). These $2p$ curves, ordered and labeled $C_1, \ldots, C_{2p}$, are supposed to issue from the point $\mathfrak{x}_0$ and to bypass the poles of the given integral. Every closed curve C on $\mathfrak{R}$ issuing from $\mathfrak{x}_0$ is homotopic to a word Q in the $4p$ letters C_k, C_k^{-1} ($k = 1, \ldots, 2p$). If g_k ($k = 1, \ldots, 2p$) is the exponent sum of C_k and C_k^{-1} in Q, where the exponents of C_k and C_k^{-1} are taken as 1 and -1 respectively, then in view of the reasoning in Section 4 of Chapter 2 (Volume I) this value depends only on C.

Theorem 2: Every period of an integral w of the first or second kind can be expressed in terms of the $2p$ fundamental periods $w(C_k)$ as follows:

$$w(C) = \sum_{k=1}^{2p} g_k w(C_k),$$

where $g_1, \ldots, g_{2p}$ are the exponent sums of the C_k in C.

Proof: Since Q arises from C by means of continuous deformation on $\mathfrak{R}$ and since the given abelian differential has no residues, the Cauchy residue theorem yields the formula $w(C) = w(Q)$, hence the assertion of the theorem.

Of particular importance in the sequel are the differentials and integrals of the first kind and their periods. In the case $p = 0$ every closed curve C is homotopic to zero, so that, by Theorem 2, $w(C) = 0$. Then an integral of the first kind $w(x)$ is single-valued on $\mathfrak{R}$, and everywhere regular, and therefore, by an argument used earlier, identically 0. In other words, in the case $p = 0$, the only differential of the first kind is 0. In the case $p = 1$, let s be the uniformizing variable. Then x and $dx/ds = y$ are elliptic functions of s, so that $ds = y^{-1}\,dx$ is a differential of the first kind. For arbitrary p we now prove

Theorem 3: Every $p + 1$ differentials of the first kind $dw_0, dw_1, \ldots, dw_p$ satisfy a nontrivial homogeneous linear equation

$$c_0\,dw_0 + c_1\,dw_1 + \cdots + c_p\,dw_p = 0$$

with constant complex coefficients.

Proof: Let

$$w_k = u_k + iv_k \qquad (k = 0, \dots, p) \tag{7}$$

be the decomposition of the integral w_k into real and imaginary parts. Using real constants ξ_k, η_k $(k = 0, \dots, p)$ we form the linear expression

$$\varphi = \sum_{k=0}^{p} (\xi_k u_k + \eta_k v_k). \tag{8}$$

If $u_k(C_l)$ and $v_k(C_l)$ are the corresponding real and imaginary part of the fundamental period $w_k(C_l)$ $(k = 0, \dots, p;\ l = 1, \dots, 2p)$, then the $2p$ homogeneous linear equations

$$\sum_{k=0}^{p} (\xi_k u_k(C_l) + \eta_k v_k(C_l)) = 0 \qquad (l = 1, \dots, 2p)$$

in the $2p + 2$ real unknowns ξ_k, η_k have a nontrivial solution. By Theorem 2 the function $\varphi(x)$ defined by (8) is single-valued on $\mathfrak{R}$. Since the w_k are regular on all of $\mathfrak{R}$, it follows that the real function $\varphi(x)$ is harmonic at all points of $\mathfrak{R}$. The compactness of $\mathfrak{R}$ implies that $\varphi(x)$ has a maximum at some point. By Theorem 6, Section 5 of Chapter 2 (Volume I), $\varphi(x)$ is therefore a constant, and, since $\varphi(x_0) = 0$, this constant is zero.

If we put

$$c_k = \xi_k - i\eta_k \qquad (k = 0, \dots, p), \qquad \sum_{k=0}^{p} c_k w_k = w,$$

then, by (7) and (8), the real part of the analytic function w is identically 0. In view of the Cauchy–Riemann equations the imaginary part of w is constant, and, since $w(x_0) = 0$, we finally obtain $w(x) = 0$. This completes the proof.

In the next section we shall prove the existence of p linearly independent differentials of the first kind $dw_1, \dots, dw_p$. Furthermore, in that section we shall construct certain special differentials of the second and third kind which we call *elementary differentials*, and define as follows. Let $\mathfrak{x}_1$ and $\mathfrak{x}_2$ be two distinct points on the Riemann surface $\mathfrak{R}$. By an *elementary differential of the third kind* on $\mathfrak{R}$ we mean a differential which is regular for $\mathfrak{x} \neq \mathfrak{x}_1$, $\mathfrak{x} \neq \mathfrak{x}_2$ and has simple poles with residues 1 and -1 at $\mathfrak{x}_1$ and $\mathfrak{x}_2$ respectively. In the next section we shall construct an elementary differential of the third kind $dE(\mathfrak{x}_1, \mathfrak{x}_2)$ for every $\mathfrak{x}_1$, $\mathfrak{x}_2$. *By an elementary differential of the second kind* we mean a differential with a single pole $\mathfrak{x}_1$ with principal part $\zeta^{-l-1}\, d\zeta$. In the next section we shall also construct an elementary differential of the second kind $dE_l(\mathfrak{x}_1)$ for every $\mathfrak{x}_1$ and arbitrary $l = 1, 2, \dots$. Here we remark that the definition of an elementary differential of the third kind is independent of the choice of the independent variable x, and that this is not the case for an elementary differential of the second kind. Hence in the latter case we suppose that we are given the Riemann surface $\mathfrak{R}$ as well as the field K. Using elementary differentials we can reduce every differential in the following manner.

Theorem 4: Let dw be a differential whose poles are at the points $\mathfrak{x}_1, \ldots, \mathfrak{x}_n$, and have orders $l_1 + 1, \ldots, l_n + 1$. If $\mathfrak{x}_0$ is a point on $\mathfrak{R}$ distinct from these poles, then there are unique complex constants a_{kl} ($l = 0, 1, \ldots, l_k$; $k = 1, \ldots, n$) and b_r ($r = 1, \ldots, p$) such that

$$dw = \sum_{k=1}^{n} \left(a_{k0}\, dE(\mathfrak{x}_k, \mathfrak{x}_0) + \sum_{l=1}^{l_k} a_{kl}\, dE_l(\mathfrak{x}_k) \right) + \sum_{r=1}^{p} b_r\, dw_r. \tag{9}$$

This decomposition is also valid for differentials of the first kind if we omit the first sum on the right.

Proof: Let c_{kl} ($l = l_k + 1, l_k, \ldots, 1$; $k = 1, \ldots, n$) be the successive coefficients of the principal part of dw at the point $\mathfrak{x}_k$. By Theorem 1,

$$\sum_{k=1}^{n} c_{k1} = 0. \tag{10}$$

If we put $a_{kl} = c_{k,l+1}$ ($l = 0, \ldots, l_k$; $k = 1, \ldots, n$), then the first sum on the right-hand side of (9) yields precisely the same principal parts at the points $\mathfrak{x}_1, \ldots, \mathfrak{x}_n$ as dw. On the other hand, (10) shows that the point $\mathfrak{x}_0$ does not contribute a pole. Thus subtraction of dw from the first sum on the right-hand side of (9) gives a differential of the first kind dw_0. Theorem 3 permits us to express dw_0 as a linear combination of $dw_1, \ldots, dw_p$. Furthermore, it is clear that the coefficients a_{kl} in (9), and, consequently, also the b_r, are uniquely determined by the principal parts at the n poles. This completes the proof.

Whereas the proof of the existence of differentials of the first kind requires additional considerations, it is easy to form differentials of the second and third kind.

Theorem 5: Let ψ be a nonconstant function from K. Then $d\psi$ is a differential of the second kind, and $d \log \psi$ is a differential of the third kind.

Proof: According to the original definition, $d\psi$ and $\psi^{-1}\, d\psi = d \log \psi$ are abelian differentials. Since the function ψ is not constant, it has at least one pole on $\mathfrak{R}$. Every pole of ψ yields a pole of next higher order of $d\psi$. The principal part of the latter has no residue, so that $d\psi$ is of the second kind. Furthermore, every pole and zero of ψ yields a nonzero residue of the differential $d \log \psi$, so that $d \log \psi$ is a differential of the third kind.

We note that, according to an earlier argument, Theorem 5 yields all the single-valued integrals on $\mathfrak{R}$.

2. *Existence*

The elementary differentials $dE_l(\mathfrak{x}_1)$ and $dE(\mathfrak{x}_0, \mathfrak{x}_1)$ are obtained with the aid of the Dirichlet principle. To this end we start from the reasoning used in Section 8 of Chapter 2 (Volume I) to construct a harmonic function with minimum property. Let $\mathfrak{R}$ be a Riemann region, $\mathfrak{p}_0$ a point of $\mathfrak{R}$, $\zeta = \xi + i\eta$ the local parameter at $\mathfrak{p}_0$ whose region of validity contains the closed disk $\mathfrak{K}_1$, $\mathfrak{K}_0$ a smaller concentric disk $|\zeta| \leqslant a$, and $\dot{\mathfrak{R}}$ the region $\mathfrak{R}$ punctured at $\mathfrak{p}_0$. We define

$$q = \frac{\xi}{\xi^2 + \eta^2} + \frac{\xi}{a^2}, \tag{1}$$

put $\hat{h} = h$ in $\mathfrak{R} - \mathfrak{K}_0$, $\hat{h} = h - q$ in $\mathfrak{K}_0$ and consider the real functions h which are continuous and piecewise differentiable on $\dot{\mathfrak{R}}$, for which $\hat{h}$ is continuous at $\mathfrak{p}_0$ and for which the Dirichlet integral $D[\hat{h};\ \mathfrak{R}]$ converges. We showed that among these functions there is one which minimizes the Dirichlet integral, and that this function u is harmonic on $\dot{\mathfrak{R}}$; whereas $u - q$ is continuous at $\mathfrak{p}_0$. The assumption of piecewise differentiability refers to a covering of $\mathfrak{R}$ with disks whose definition we shall not repeat. We had then made the simplifying assumptions that $\mathfrak{p}_0$ is not a branch point or a point at infinity and that it lies over $z = 0$, so that $\zeta = z$. It is easy to see that these assumptions are not essential. In the proof it was important that the normal derivative of q vanishes on the boundary C of $\mathfrak{K}_0$, and that, for $\zeta \neq 0$, q is harmonic on the disk $\mathfrak{K}_1$ containing in its interior the disk $\mathfrak{K}_0$. If we bear in mind these properties of the function defined by (1), then we can generalize the statement of the extremum problem as follows. All we now assume of the real-valued function q is that it is harmonic on a subdomain of $\mathfrak{R}$ containing, in addition to the boundary C, the ring $\mathfrak{K}_1 - \mathfrak{K}_0$, and that its normal derivative vanishes throughout C. We shall call such a function a *singularity function*, for it may have discontinuities in the interior of $\mathfrak{K}_0$, or may even fail to be defined on a disk concentric with and smaller than $\mathfrak{K}_0$.

The definition of the comparison function h and of $\hat{h}$ should be modified as follows. The function need be defined only on $\mathfrak{R} - \mathfrak{K}_0$, and is supposed continuous and piecewise differentiable there. Furthermore, $\hat{h}$ is any function continuous on $\mathfrak{K}_0$ and continuously differentiable in its interior which coincides with $h - q$ on C. As before, we require the convergence of

$$D[\hat{h};\ \mathfrak{R}] = D[h;\ \mathfrak{R} - \mathfrak{K}_0] + D[\hat{h};\ \mathfrak{K}_0].$$

The original proof carries over in a natural way to the case of a singularity function and yields the existence of a harmonic function u on $\mathfrak{R} - \mathfrak{K}_0$ such that the harmonic function $u^* = u - q$ defined on $\mathfrak{R}_1 - \mathfrak{R}_0$ can be harmonically continued to all points of $\mathfrak{R}_0$, and so yields a harmonic function throughout $\mathfrak{R}_1$. The fact that the Dirichlet integral is minimized for $h = u$, $\hat{h} = u^*$ is of no consequence in the sequel. Since the earlier proof applies with hardly any modification to the case of an arbitrary singularity function, and no new difficulties arise, we simply leave the matter at that. When applying the existence theorem we suppose the Riemann region $\mathfrak{R}$ compact, and always define the singularity function q in a very simple manner.

Theorem 1: For an arbitrary point $\mathfrak{x}_1$ on $\mathfrak{R}$ and $l = 1, 2, \ldots$ there exist abelian differentials of the second kind $dE_l(\mathfrak{x}_1)$ and $dF_l(\mathfrak{x}_1)$ with a single pole at $\mathfrak{x}_1$ with principal part $\zeta^{-l-1}\, d\zeta$. Furthermore, all the periods of the integral $E_l(\mathfrak{x}_1)$ are pure imaginary and those of $F_l(\mathfrak{x}_1)$ are real.

Proof: We apply the existence proof discussed above to the case when $\mathfrak{p}_0 = \mathfrak{x}_1$ and take for the singularity function q, as generalization of (1), the real part of the function

$$g = g(\zeta) = \zeta^{-l} + a^{-2l}\zeta^{l} \qquad (l = 1, 2, \ldots).$$

If we introduce polar coordinates on $\mathfrak{R}_1$ by means of the substitution $\zeta = re^{i\varphi}$, then

$$q = \frac{g + \bar{g}}{2} = (r^{-l} + a^{-2l}r^{l}) \cos(l\varphi);$$

and, from

$$q_r = \frac{l}{r}(a^{-2l}r^{l} - r^{-l}) \cos(l\varphi),$$

we obtain for the normal derivative on the boundary $r = a$ of $\mathfrak{R}_0$, the value 0, as required for application of the existence theorem. Hence for every $l = 1, 2, \ldots$, there exists on $\mathfrak{R}$ a function u harmonic for $\mathfrak{x} = \mathfrak{x}_1$ such that the difference $u - q$ is harmonic at $\mathfrak{x}_1$ as well. For $l = 1$ this was proved in great detail in Section 8 of Chapter 2 (Volume I).

The conjugate harmonic function v of u is determined in accordance with the Cauchy–Riemann equations from the formula

$$dv = v_\xi\, d\xi + v_\eta\, d\eta = u_\xi\, d\eta - u_\eta\, d\xi.$$

For the meromorphic function $w = u + iv$ we have

$$\begin{aligned} dw &= d(u + iv) = u_\xi\, d\xi + u_\eta\, d\eta + i(u_\xi\, d\eta - u_\eta\, d\xi) \\ &= (u_\xi - iu_\eta)\, d\zeta. \end{aligned} \tag{2}$$

Similarly,

$$dg = d(\zeta^{-l} + a^{-2l}\zeta^{l}) = (q_\xi - iq_\eta)\, d\zeta,$$

so that $w - g$ remains regular at $\mathfrak{x}_1$, and dw has there the principal part

$$d\zeta^{-l} = -l\zeta^{-l-1}\,d\zeta.$$

If x is again the independent variable, then, in view of (2) and the properties of u, the function

$$\frac{dw}{dx} = \frac{dw}{d\zeta}\bigg/\frac{dx}{d\zeta}$$

is single-valued and meromorphic on $\mathfrak{R}$, that is, belongs to K. Hence dw is seen to be an abelian differential of the second kind whose only pole is at $\mathfrak{x}_1$, and the principal part at that pole is $-l\zeta^{-l-1}\,d\zeta$. Furthermore, the real part u of the function w on $\mathfrak{R}$ is single-valued. It follows that the differential

$$dE_l(\mathfrak{x}_1) = -\frac{1}{l}\,dw \qquad (l = 1, 2, \ldots)$$

has the required properties.

Similarly, to obtain the abelian integral $F_l(\mathfrak{x}_1)$ we define q to be the imaginary part of the function $a^{-2l}\zeta^l - \zeta^{-l}$, that is, we put

$$q = (r^{-l} + a^{-2l}r^l)\sin(l\varphi),$$

and otherwise proceed just as above in the determination of w. In the present case the imaginary part of iw is single-valued on $\mathfrak{R}$, and

$$dF_l(\mathfrak{x}_1) = \frac{i}{l}\,dw$$

yields the needed differential.

It is easy to see that the differentials $dE_l(\mathfrak{x}_1)$ and $dF_l(\mathfrak{x}_1)$ are uniquely determined by the properties listed in Theorem 1. In fact, the difference of two such differentials leads to an integral of the first kind whose real or imaginary part is single-valued and harmonic on $\mathfrak{R}$, and must therefore be 0.

Theorem 2: For any two given distinct points $\mathfrak{x}_1$ and $\mathfrak{x}_2$ on $\mathfrak{R}$ there exists a differential $dE(\mathfrak{x}_1, \mathfrak{x}_2)$ of the third kind which is regular apart from simple poles at $\mathfrak{x}_1$ and $\mathfrak{x}_2$ with residues 1 and -1. Also, all the periods of the integral $E(\mathfrak{x}_1, \mathfrak{x}_2)$ are pure imaginary.

Proof: We choose $\mathfrak{p}_0 = \mathfrak{x}_1$, and suppose at first that $\mathfrak{x}_2$ also lies in $\mathfrak{K}_0$. Then the values of the local parameter at the points $\mathfrak{x}_1$ and $\mathfrak{x}_2$ are 0 and ζ_2, where $0 < |\zeta_2| < a$. Let ζ_3 be the reflection of ζ_2 in the boundary C of $\mathfrak{K}_0$, that is,

$$\zeta_3\bar{\zeta}_2 = a^2, \qquad |\zeta_3| > a.$$

We may suppose $\mathfrak{K}_1$ constructed so that ζ_3 also lies outside $\mathfrak{K}_1$ (Figure 34). Every branch of the function

$$g(\zeta) = \log\frac{\zeta}{(\zeta - \zeta_2)(\zeta - \zeta_3)} \tag{3}$$

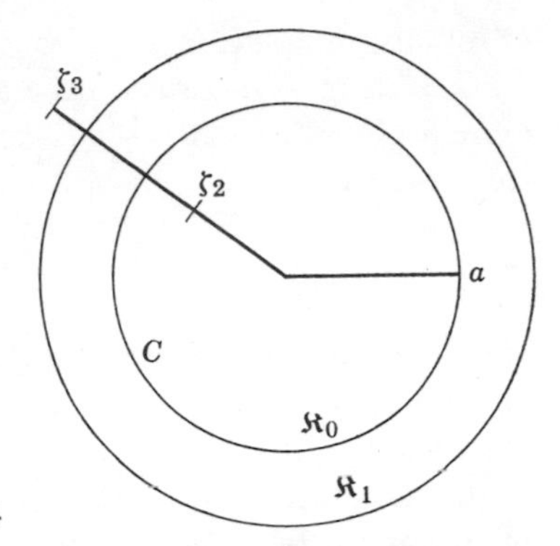

Figure 34

is single-valued and regular on the ring $\mathfrak{K}_1 - \mathfrak{K}_0$ and its boundary. We can then define $q = \frac{1}{2}(g + \bar{g})$ as the real part of $g(\zeta)$, provided that we can verify the vanishing of the radial derivative on C. If we introduce polar coordinates by means of the substitution $\zeta = re^{i\varphi}$, then it follows from the Cauchy–Riemann differential equations that the derivative of q with respect to $\log r$ is equal to the derivative of the imaginary part of $g(\zeta)$ with respect to φ. Since, for $\zeta\bar{\zeta} = a^2$, the ratio

$$\frac{\bar{\zeta}}{(\bar{\zeta} - \bar{\zeta}_2)(\bar{\zeta} - \bar{\zeta}_3)} \bigg/ \frac{\zeta}{(\zeta - \zeta_2)(\zeta - \zeta_3)} = \zeta_2/\bar{\zeta}_2$$

is independent of φ it follows that $q_r = 0$.

By the existence theorem there exists a function u harmonic on $\mathfrak{R} - \mathfrak{K}_0$ such that the difference $u - q$ is harmonic throughout $\mathfrak{K}_1$. If v is the harmonic conjugate of u, it follows as in the proof of Theorem 1 that, with $w = u + iv$, dw is an abelian differential of the third kind. Since ζ_3 is outside $\mathfrak{K}_1$, dw is regular on $\mathfrak{R}$ except at the points $\mathfrak{x}_1$ and $\mathfrak{x}_2$, where it has simple poles at which the residues, by (3), are 1 and -1. Here we note that, at any rate, the local parameter at $\mathfrak{x}_2$ has the form

$$\tau = \zeta^j - \zeta_2^j$$

with integer $j \neq 0$, and

$$-d \log (\zeta - \zeta_2)^{-1} = d \log \{\zeta_2(1 + \zeta_2^{-j}\tau)^{1/j} - \zeta_2\} = (\tau^{-1} + \cdots)\, d\tau.$$

Since, in addition, the real part u of w is single-valued on all of $\mathfrak{R}$, all the periods of w are pure imaginary. It follows that the differential

$$dE(\mathfrak{x}_1, \mathfrak{x}_2) = dw$$

has all the required properties.

If the point $\mathfrak{x}_2$ is not inside $\mathfrak{K}_0$, then we join $\mathfrak{x}_1$ to $\mathfrak{x}_2$ by means of a curve on $\mathfrak{R}$. By introducing a sufficiently fine subdivision we can find a succession of points $\mathfrak{p}_0 = \mathfrak{x}_1, \mathfrak{p}_1, \ldots, \mathfrak{p}_n = \mathfrak{x}_2$ such that every pair of points $\mathfrak{p}_{k-1}$, $\mathfrak{p}_k$ $(k = 1, \ldots, n)$ satisfies the assumptions on $\mathfrak{x}_1$ and $\mathfrak{x}_2$. This implies the existence of the n differentials of the third kind $dE(\mathfrak{p}_{k-1}, \mathfrak{p}_k)$, so that we can form the sum

$$dE(\mathfrak{x}_1, \mathfrak{x}_2) = \sum_{k=1}^{n} dE(\mathfrak{p}_{k-1}, \mathfrak{p}_k). \tag{4}$$

Here the poles cancel out with the exception of the poles at $\mathfrak{p}_0 = \mathfrak{x}_1$ and $\mathfrak{p}_n = \mathfrak{x}_2$ with residues 1 and -1. It follows that the differential (4) has all the required properties.

Just as before we see that $dE(\mathfrak{x}_1, \mathfrak{x}_2)$ is uniquely determined by the conditions of Theorem 2. $dE_l(\mathfrak{x}_1)$ $(l = 1, 2, \ldots)$ and $dE(\mathfrak{x}_1, \mathfrak{x}_2)$ yield the elementary differentials of the second and third kind which we used already in the preceding section. To obtain p linearly independent differentials of the first kind we must first construct the differentials of the third kind which correspond to the differentials of the second kind denoted by $dF_l(\mathfrak{x}_1)$.

Theorem 3: Let $\mathfrak{x}_1$ and $\mathfrak{x}_2$ be two distinct points on $\mathfrak{R}$ connected by a simple, oriented curve L, and let $\mathfrak{R}^*$ be the region obtained by cutting along L. There exists a differential of the third kind $dF(\mathfrak{x}_1, \mathfrak{x}_2)$ which is regular in $\mathfrak{R}$ apart from simple poles at $\mathfrak{x}_1$ and $\mathfrak{x}_2$ with residues 1 and $-$ 1. The periods of $F(\mathfrak{x}_1, \mathfrak{x}_2)$ on $\mathfrak{R}^*$ are all real, and, for every loop which crosses L exactly once from right to left, the imaginary part of the period is 2π.

Proof: We proceed in analogy to the proof of Theorem 2. With $\mathfrak{p}_0$, ζ_2, ζ_3, defined as before, we introduce the function

$$g(\zeta) = \log \frac{\zeta(\zeta - \zeta_3)}{(\zeta - \zeta_2)}, \tag{5}$$

choose on $\mathfrak{K}_1 - \mathfrak{K}_0$ a definite, and therefore single-valued, branch, and take as singularity function its imaginary part

$$q = \frac{g - \bar{g}}{2i}.$$

If we once more put $\zeta = re^{i\phi}$ then, this time, for $r = a$, the product

$$\frac{\zeta(\zeta - \zeta_3)}{\zeta - \zeta_2} \cdot \frac{\bar{\zeta}(\bar{\zeta} - \bar{\zeta}_3)}{\bar{\zeta} - \bar{\zeta}_2} = \zeta_3\bar{\zeta}_3$$

is independent of φ; so that the derivative of the real part of $g(\zeta)$ with respect to φ vanishes there, and we again have $q_r = 0$ on the boundary of $\mathfrak{K}_0$.

By the existence theorem there is a function v harmonic on $\mathfrak{R} - \mathfrak{K}_0$ and such that the difference $v - q$ is harmonic on $\mathfrak{K}_1$. If we form $w = u + iv$

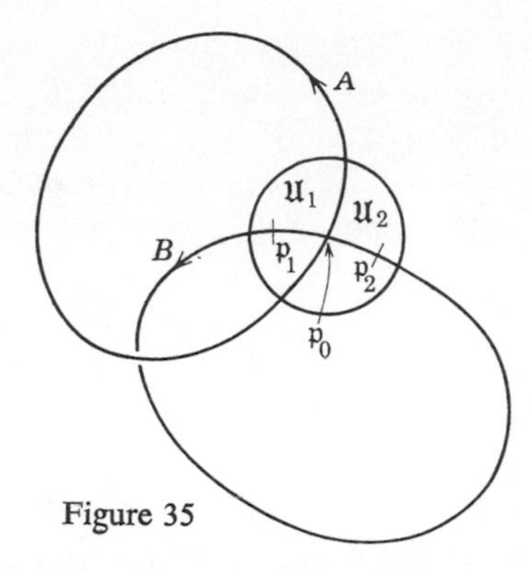

Figure 35

with u the harmonic conjugate of v, then $w - g(\zeta)$ is regular on $\mathfrak{K}_1$, and

$$dF(\mathfrak{x}_1, \mathfrak{x}_2) = dw$$

is seen to be a differential of the third kind which is regular everywhere with the exception of the simple poles at $\mathfrak{x}_1$ and $\mathfrak{x}_2$ with residues 1 and -1. If we cut in the interior along a simple curve from $\mathfrak{x}_1$ to $\mathfrak{x}_2$, then, by (5) and (6), v is single-valued on $\mathfrak{R}^*$, and upon crossing L it jumps by 2π provided that L is crossed from right to left. This implies the desired conclusion in the case when $\mathfrak{x}_1 = \mathfrak{p}_0$ and L lies in the interior of $\mathfrak{K}_0$. In the general case $dF(\mathfrak{x}_1, \mathfrak{x}_2)$ is obtained as an additive combination in analogy to (4). This completes the proof.

Let A be a simple, closed curve on $\mathfrak{R}$ which does not separate $\mathfrak{R}$ into two disjoint parts. In the case $p = 0$, $\mathfrak{R}$ is homeomorphic to a sphere and, by Jordan's theorem, the existence of such a curve is ruled out. We choose a fixed point $\mathfrak{p}_0$ on A. The curve A separates a sufficiently small neighborhood $\mathfrak{U}$ of $\mathfrak{p}_0$ into two disjoint parts $\mathfrak{U}_1$ and $\mathfrak{U}_2$. We choose a point $\mathfrak{p}_1$ in $\mathfrak{U}_1$, and a point $\mathfrak{p}_2$ in $\mathfrak{U}_2$, and join them on $\mathfrak{R}$ by means of a simple curve B^* which does not meet A; this is possible, for cutting $\mathfrak{R}$ along A does not disconnect it. Now we join $\mathfrak{p}_1$ to $\mathfrak{p}_0$ on $\mathfrak{U}_1$ and $\mathfrak{p}_2$ to $\mathfrak{p}_0$ on $\mathfrak{U}_2$ in an appropriate manner and complete B^* to a simple, closed curve B which meets the curve A only at the point $\mathfrak{p}_0$, and crosses it at that point. With A oriented, the orientation of B can be chosen so that B crosses the curve A at $\mathfrak{p}_0$ from right to left, that is, in the manner of the imaginary axis crossing the real axis (Figure 35). Then A and B are called a pair of *conjugate crosscuts*. We note that B is conjugate with A^{-1}.

Theorem 4: Let A, B be a pair of conjugate crosscuts on $\mathfrak{R}$ and let $\mathfrak{R}^*$ be the domain obtained by cutting $\mathfrak{R}$ along A. Then there exists exactly one differential dw of the first kind such that the period $w(B)$ has real part 1, and for all closed paths C on $\mathfrak{R}^*$ the period $w(C)$ is pure imaginary.

Proof: Let $\mathfrak{x}_1$ and $\mathfrak{x}_2$ be two points on A which do not lie at the point of intersection $\mathfrak{p}_0$ of A and B. We denote by L and M the directed subarcs of A determined by $\mathfrak{x}_1$ and $\mathfrak{x}_2$ so that L leads from $\mathfrak{x}_1$ to $\mathfrak{x}_2$ and M leads from $\mathfrak{x}_2$ to $\mathfrak{x}_1$. We apply Theorem 3 with $\mathfrak{x}_1$, $\mathfrak{x}_2$, L and $\mathfrak{x}_2$, $\mathfrak{x}_1$, M. Adding the corresponding differentials of the third kind

$$\frac{1}{2\pi i}\, dF(\mathfrak{x}_1, \mathfrak{x}_2) = dF_L, \frac{1}{2\pi i}\, dF(\mathfrak{x}_2, \mathfrak{x}_1) = dF_M,$$

we obtain the abelian differential

$$dw = dF_L + dF_M.$$

This differential is of the third kind, for, as a result of addition, the only poles $\mathfrak{x}_1$ and $\mathfrak{x}_2$ of the two summands drop out. By Theorem 3, the periods of the integral w have the required properties; it should be noted that B crosses L or M. Subtraction of two differentials of the required kind must, by a frequently used argument, yield 0, which proves the uniqueness part of our assertion.

Theorem 5: If $\mathfrak{R}$ is of genus p then there are p linearly independent differentials of the first kind.

Proof: We start with a canonical dissection of $\mathfrak{R}$ by means of p pairs of conjugate crosscuts A_k, B_k $(k = 1, \ldots, p)$. All of them issue from the same point $\mathfrak{p}$ and are otherwise disjoint. In view of the construction in Section 2 of Chapter 2 (Volume I), the curves A_k, B_k $(k = 1, \ldots, p)$ are not crossed at $\mathfrak{p}$ by the remaining $2p - 2$ curves A_l, B_l $(l \neq k)$. Let us temporarily denote the $2p$ curves A_k, B_k $(k = 1, \ldots, p)$ by $C_1, \ldots, C_{2p}$ with $A_k = C_k$ and $B_k = C_{k+p}$. For $k = 1, \ldots, 2p$ let $\mathfrak{R}_k$ be the result of cutting $\mathfrak{R}$ along the crosscut conjugate to C_k. By means of Theorem 4 we construct $2p$ differentials dw_k $(k = 1, \ldots, 2p)$ of the first kind such that the real part of the period $w_k(C_k)$ is 1 and, for all closed paths C on $\mathfrak{R}_k$, the value $w(C)$ is pure imaginary. Now for $l \neq k$, C_l lies on $\mathfrak{R}_k$, for the crosscut conjugate to C_k yields the boundary of $\mathfrak{R}_k$, and that is not crossed by C_l at $\mathfrak{p}$. It follows that for $l \neq k$, $w_k(C_l)$ is pure imaginary and, consequently, the matrix whose entries are the real parts of the periods $w_k(C_l)$ $(k, l = 1, \ldots, 2p)$ is the unit matrix of order $2p$.

Using any real numbers $\lambda_1, \ldots, \lambda_{2p}$ we form the differential of the first kind

$$dw = \lambda_1\, dw_1 + \cdots + \lambda_{2p}\, dw_{2p}.$$

Then, clearly, λ_k $(k = 1, \ldots, 2p)$ is the real part of the period $w(C_k)$. Hence $dw = 0$ only in the trivial case $\lambda_1 = 0, \ldots, \lambda_{2p} = 0$, which proves the linear independence of the $2p$ differentials dw_k $(k = 1, \ldots, 2p)$ over the reals. Now let q be the complex rank of the dw_k. This means that every $q + 1$ of these differentials are linearly dependent over the complex numbers, and some q of these differentials, to be denoted by $\omega_1, \ldots, \omega_q$, are linearly independent over the complex numbers. It is clear that the $2q$ differentials $\omega_k, i\omega_k$ $(k = 1, \ldots, q)$ are linearly independent over the reals and that every dw_k is linearly dependent on them over the reals. Since $2q$ homogeneous linear equations in $2p$ unknowns with real coefficients have in the case $q < p$ a nontrivial real solution, it follows that $q \geqslant p$. By Theorem 3, Section 1, we have on the other hand $q \leqslant p$. It follows that $q = p$, and, after a possible change of indices, we can claim that $dw_1, \ldots, dw_p$ are linearly independent over the complex numbers. Every system of p linearly independent differentials of the first kind can be obtained from the latter system by means of a linear transformation with complex coefficients whose determinant is different from 0.

In the preceding discussion the differentials of the third kind were particularly important, since these differentials enabled us to construct the differentials of the first kind. Without going into the matter, we wish to point out that it is possible to construct a basis of p differentials of the first kind by means of a purely algebraic procedure using the properties of the field K, and thus give a purely algebraic definition of the number p.

3. *The period matrix*

In this and the next two sections we investigate more closely the abelian integrals of the first kind, and, in particular, their periods. Since there are no nontrivial differentials of the first kind when $p = 0$, we assume for the time being that $p > 0$. We start again with a canonical dissection of $\mathfrak{R}$ with pairs of crosscuts A_k, B_k $(k = 1, \ldots, p)$. As a result of this dissection the Riemann surface goes over into a simply connected domain $\mathfrak{R}^*$ whose boundary is the positively oriented curve

$$C = T_1 T_2 \cdots T_p, \qquad T_k = A_k B_k A_k^{-1} B_k^{-1} \qquad (k = 1, \ldots, p). \tag{1}$$

Theorem 1: If φ and ψ are two arbitrary integrals of the first kind, then

$$\sum_{k=1}^{p} (\varphi(A_k)\psi(B_k) - \varphi(B_k)\psi(A_k)) = 0.$$

Proof: The functions φ and ψ are regular throughout $\mathfrak{R}^*$, and, by the monodromy theorem, single-valued. By the introduction of a global uniformizing parameter t, $\mathfrak{R}^*$ is mapped onto a schlicht domain $\mathfrak{F}$ in the plane with a simple, closed boundary curve L. By Cauchy's theorem,

$$\int_C \varphi \, d\psi = \int_L \varphi \frac{d\psi}{dt} \, dt = 0. \tag{2}$$

We consider the values of φ on the boundary C of $\mathfrak{R}^*$. If we go from a point $\mathfrak{p}$ on the portion A_k of the boundary along any path in $\mathfrak{R}^*$ to the point $\mathfrak{p}^*$ corresponding to $\mathfrak{p}$ on A_k^{-1}, then, by (1), the value of φ is increased by the period $\varphi(B_k)$. After the corresponding transition from B_k to B_k^{-1} the value of φ is increased by the period $\varphi(A_k^{-1}) = -\varphi(A_k)$. In both cases the differential $d\psi$ remains unchanged (Figure 36). Since A_k and A_k^{-1} as well as B_k and B_k^{-1} are oppositely oriented, it follows that

$$(3)\quad \int_C \varphi\, d\psi = \sum_{k=1}^{p}\left(\int_{A_k}\{\varphi - (\varphi + \varphi(B_k))\}\, d\psi + \int_{B_k}\{\varphi - (\varphi - \varphi(A_k))\}\, d\psi\right)$$

$$= \sum_{k=1}^{p}\left(\varphi(A_k)\int_{B_k} d\psi - \varphi(B_k)\int_{A_k} d\psi\right) = \sum_{k=1}^{p}(\varphi(A_k)\psi(B_k) - \varphi(B_k)\psi(A_k)).$$

By (2), this implies our assertion.

For $p = 1$ the theorem is trivial, since, in that case, φ and ψ differ only by a constant factor.

We denote by $[\varphi]$ the column of the $2p$ values $\varphi(A_1), \ldots, \varphi(A_p)$, $\varphi(B_1), \ldots, \varphi(B_p)$ taken in this order. We let E denote the $p \times p$ unit matrix, and we introduce the $2p \times 2p$ skew-symmetric matrix

$$(4)\qquad J = \begin{pmatrix} 0 & E \\ -E & 0 \end{pmatrix}.$$

Letting $[\varphi]'$ be the row obtained by transposing the column $[\varphi]$ we can state Theorem 1 in the simple form

$$(5)\qquad [\varphi]'J[\psi] = 0.$$

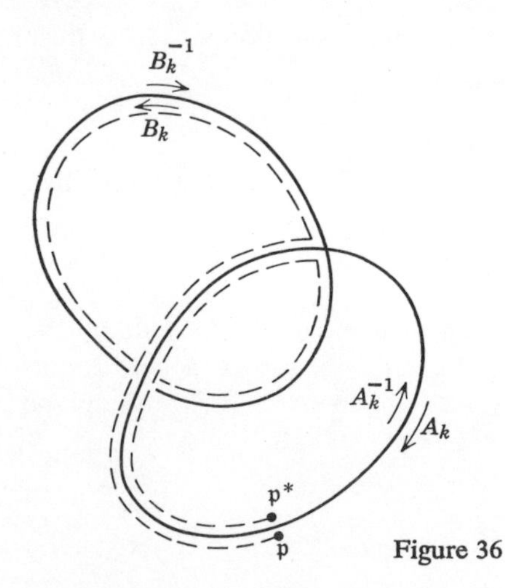

Figure 36

If $w_1, \ldots, w_p$ are any p linearly independent integrals of the first kind and as such form a basis for all integrals of the first kind, then we form the matrix

$$Q = ([w_1][w_2] \cdots [w_p])$$

with p columns and $2p$ rows. We call this matrix the *period matrix.* This matrix depends, in addition to the choice of basis, on the initial canonical dissection of $\mathfrak{R}$. If we choose $\varphi = w_k$, $\psi = w_l$ and put for k and l the values $1, \ldots, p$, then (5) implies the matrix equality

$$Q'JQ = 0. \tag{6}$$

Since the matrix on the left-hand side of (6) is skew-symmetric, we obtain from (6) altogether $p(p-1)/2$ different homogeneous equations of degree two for the $2p$ periods of all the basis elements $w_1, \ldots, w_p$.

Theorem 2: For any integral of the first kind φ which is not identically zero we have the inequality

$$\frac{1}{2i}\sum_{k=1}^{p}(\bar{\varphi}(A_k)\varphi(B_k) - \bar{\varphi}(B_k)\varphi(A_k)) > 0. \tag{7}$$

Proof: Letting $\varphi = u + iv$, we have

$$\bar{\varphi}\, d\varphi = (u - iv)(du + i\, dv) = d\left(\frac{u^2 + v^2}{2}\right) + i(u\, dv - v\, du).$$

In view of the fact that $u^2 + v^2$ is single-valued on the simply connected surface $\mathfrak{R}^*$, we have

$$\frac{1}{2i}\int_C \bar{\varphi}\, d\varphi = \tfrac{1}{2}\int_C u\, dv - v\, du = \iint_{\mathfrak{B}} du\, dv,$$

where $\mathfrak{B}$ is the not necessarily schlicht image of $\mathfrak{R}^*$ over the w-plane given by the function $w = \varphi$. If we again introduce the global uniformizing parameter $t = r + is$, then the functional determinant

$$\frac{d(u, v)}{d(r, s)} = u_r v_s - v_r u_s = u_r^2 + v_r^2 = \left|\frac{d\varphi}{dt}\right|^2.$$

Hence

$$\iint_{\mathfrak{B}} du\, dv = \iint_{\mathfrak{F}} \left|\frac{d\varphi}{dt}\right|^2 dr\, ds > 0,$$

since the derivative of φ with respect to t is not identically 0. This proves the inequality

$$\frac{1}{2i}\int_C \bar{\varphi}\, d\varphi > 0. \tag{8}$$

Now the left-hand side of (8) can be subjected to the transformation carried out in (3) with φ and ψ replaced by $\bar{\varphi}$ and φ. The result is the assertion in (7). We note that the double integral under consideration is precisely the Dirichlet integral $D[u;\ \mathfrak{R}]$.

We apply Theorem 2 to

$$\varphi = z_1 w_1 + \cdots + z_p w_p,$$

with $z_1, \ldots, z_p$ arbitrary complex coefficients not all zero. Since $w_1, \ldots, w_p$ are linearly independent integrals of the first kind, φ is not identically 0, and the relation (7) holds. This relation can be written as

$$\frac{1}{2i}[\bar{\varphi}]'J[\varphi] > 0.$$

If z is the column with entries $z_1, \ldots, z_p$, then

$$[\varphi] = Qz,$$

and we obtain the inequality

$$\frac{1}{2i}\bar{z}'\bar{Q}'JQz > 0 \qquad (z \neq 0). \tag{9}$$

The matrix

$$H = \frac{1}{2i}\bar{Q}'JQ$$

has the property

$$H' = \frac{1}{2i}Q'J'\bar{Q} = -\frac{1}{2i}\overline{Q'JQ} = \bar{H}$$

and is therefore hermitian. Hence inequality (9) states that the hermitian form

$$\sum_{k,l=1}^{p} h_{kl}\bar{z}_k z_l = \bar{z}'Hz$$

is positive definite, that is, it takes on positive real values for all complex numbers $z \neq 0$. This property is expressed by writing $H > 0$. Here we make the following observation. If a hermitian matrix H is real, then it is symmetric; the condition $H > 0$ is, in this case, equivalent to the requirement that the quadratic form $z'Hz$ is positive for all real columns $z \neq 0$, that is, that the form is positive definite.

Noting (6), we have the following two results pertaining to the period matrix Q:

$$Q'JQ = 0, \qquad \frac{1}{2i}\bar{Q}'JQ > 0. \tag{10}$$

These relations were found in the same way by Riemann and are called the *Riemann period relations*. It is shown in algebra that a hermitian form $z'Hz$ in p variables $z_1, \ldots, z_p$ is positive definite if and only if for $k = 1, \ldots, p$ the real determinant formed from the first k rows and columns of H is positive. This adds p inequalities to the $p(p-1)/2$ equalities already established for the period matrix.

In the case $p = 1$,

$$Q = \begin{pmatrix} \omega_1 \\ \omega_2 \end{pmatrix}$$

is the column whose entries are the two fundamental periods of an elliptic integral of the first kind. In this case equality (6) holds identically in ω_1 and ω_2, while inequality (9) yields the condition

$$\frac{1}{2i}\bar{Q}'JQ = \frac{1}{2i}(\bar{\omega}_1 \quad \bar{\omega}_2)\begin{pmatrix} 0 & 1 \\ -1 & 0 \end{pmatrix}\begin{pmatrix} \omega_1 \\ \omega_2 \end{pmatrix} = \frac{\bar{\omega}_1\omega_2 - \bar{\omega}_2\omega_1}{2i} > 0. \tag{11}$$

This condition states that the ratio ω_2/ω_1 of the periods is not real and that the vectors ω_1, ω_2 are disposed in the manner of the real and imaginary axes. Comparison with Section 10 of Chapter 1 (Volume I) shows that our proof of Theorem 2 generalizes an idea used in that section.

We now wish to extend the formalism of inequality (11) to the case of arbitrary p. Since Q consists of $2p$ rows and p columns, we can write

$$Q = \begin{pmatrix} F \\ G \end{pmatrix},$$

where F and G are $p \times p$ matrices. We then have, by (10),

$$Q'JQ = (F'G')\begin{pmatrix} 0 & E \\ -E & 0 \end{pmatrix}\begin{pmatrix} F \\ G \end{pmatrix} = F'G - G'F = 0, \tag{12}$$

$$H = \frac{1}{2i}\bar{Q}'JQ = \frac{1}{2i}(\bar{F}'\bar{G}')\begin{pmatrix} 0 & E \\ -E & 0 \end{pmatrix}\begin{pmatrix} F \\ G \end{pmatrix} = \frac{1}{2i}(\bar{F}'G - \bar{G}'F) > 0. \tag{13}$$

Theorem 3: For every period matrix Q the determinant $|F|$ is different from 0, the matrix $GF^{-1} = Z = X + iY$ is symmetric, and its imaginary part Y is >0.

Proof: Let the column s of complex numbers be a solution of the equation

$$Fs = 0. \tag{14}$$

Then, by (13)

$$\bar{s}'Hs = \frac{1}{2i}\{\overline{(Fs)}'Gs - \overline{(Gs)}'Fs\} = 0.$$

Since $H > 0$, $s = 0$. Hence (14) has only the trivial solution $s = 0$, and it follows that $|F| \neq 0$. Multiplying (12) on the left by $(F')^{-1}$ and on the right by F^{-1} we obtain for the matrix $Z = GF^{-1}$ the relation

$$Z - Z' = 0.$$

Thus $Z = X + iY$ is a symmetric matrix. Furthermore, the hermitian matrix $(\bar{F}')^{-1}HF^{-1}$ is also positive definite, and, on the other hand, by (13)

$$(\bar{F}')^{-1}HF^{-1} = \frac{1}{2i}(GF^{-1} - (\bar{F}')^{-1}\bar{G}') = \frac{1}{2i}(Z - \bar{Z}') = \frac{1}{2i}(Z - Z') = Y$$

is real symmetric. This proves all the assertions of the theorem.

If $w_1, \ldots, w_p$ is a given basis of integrals of the first kind, then all other bases are obtained from it by forming the row $(w_1, \ldots, w_p)T$, where T varies over all the $p \times p$ complex matrices with $|T| \neq 0$. It follows that the period matrix of the transformed matrix is

$$([w_1][w_2] \cdots [w_p])T = QT = \begin{pmatrix} FT \\ GT \end{pmatrix}.$$

Here T can be uniquely determined by requiring $FT = E$, that is, $T = F^{-1}$; then $GT = GF^{-1} = Z$. Given a canonical dissection of $\mathfrak{R}$, this determines precisely one basis for the abelian integrals of the first kind, which we call a *normal basis* and again denote by $w_1, \ldots, w_p$. Then the associated period matrix

$$Q = \begin{pmatrix} E \\ Z \end{pmatrix}$$

is said to be *normalized*. Hence a normalized period matrix is characterized by the condition $F = E$. We shall also refer briefly to the second component Z as the *period matrix* provided the meaning is clear from the context. The period matrix Z is symmetric of order p and has a positive imaginary part Y.

In the case $p = 1$, normalization of the period matrix Q means that we impose on the, previously arbitrary, constant factor in the elliptic integral of the first kind the condition that one fundamental period has the value 1. As a result the other period has a positive imaginary part.

4. The modular group

It is important to investigate the nature of the dependence of the period matrix on the initial canonical dissection of the Riemann surface $\mathfrak{R}$. In this connection the modular group of degree p will play a special role. To define that group we first introduce the so-called symplectic group.

A real matrix L of order $2p$ is called *symplectic* if it satisfies the equation

$$L'JL = J, \tag{1}$$

where J is the matrix defined by formula (4) of the preceding section. Since $|J| = 1 \neq 0$, (1) implies that $|L|^2 = 1$, that is, $|L| = \pm 1 \neq 0$. More precisely, it is possible to show that we always have $|L| = 1$; we shall make no use of this result in the sequel. At any rate, L^{-1} exists and we see from (1) that it is symplectic. Since the product L_1L_2 of symplectic matrices L_1 and L_2 is a symplectic matrix, it follows that the symplectic matrices form a group under multiplication. The relation $J^{-1} = -J$ and definition (1) show that the transpose L' of a symplectic matrix L is also symplectic. If we write L in the form of a block matrix

$$L = \begin{pmatrix} A & B \\ C & D \end{pmatrix},$$

where A, B, C, D are submatrices of order p, then (1) goes over into

$$\begin{pmatrix} 0 & E \\ -E & 0 \end{pmatrix} = J = L'JL = \begin{pmatrix} A' & C' \\ B' & D' \end{pmatrix}\begin{pmatrix} 0 & E \\ -E & 0 \end{pmatrix}\begin{pmatrix} A & B \\ C & D \end{pmatrix}$$
$$= \begin{pmatrix} A'C - C'A & A'D - C'B \\ B'C - D'A & B'D - D'B \end{pmatrix},$$

so that

(2)

$$A'C = C'A, \qquad A'D - C'B = E, \qquad B'C - D'A = -E, \qquad B'D = D'B.$$

The third of these equalities can be obtained from the second by multiplication by -1 and subsequent transposition. The remaining two equalities state that $A'C$ and $B'D$ are both symmetric. Now L is symplectic if and only if L' is symplectic. It follows that conditions (2) are equivalent to the conditions

$$AB' = BA', \qquad CD' = DC', \qquad AD' - BC' = E. \tag{3}$$

With a view to future applications we write conditions (2) in yet another form. If x_1, x_2, y_1, y_2 are four columns of p elements each, then we put

$$x = \begin{pmatrix} x_1 \\ x_2 \end{pmatrix}, \qquad y = \begin{pmatrix} y_1 \\ y_2 \end{pmatrix} \tag{4}$$

and call the expression

$$x'Jy = x_1'y_2 - x_2'y_1$$

the *alternating product* of x and y. Using (2), or directly from (1), we obtain the following conditions for the $2p$ successive columns $l_1, l_2, \ldots, l_{2p}$ of L:

$$l'_g J l_h = 0 \ (g - h \neq \pm p;\ g, h = 1, \ldots, 2p) \tag{5}$$
$$l'_g J l_{g+p} = 1 \ (g = 1, \ldots, p).$$

For $p = 1$,

$$L = \begin{pmatrix} a & b \\ c & d \end{pmatrix}$$

is of order two, and the conditions (5) state that the determinant $ad - bc$ has the value 1.

If A, B, C, D are integral matrices, that is, all their entries are integers, then the symplectic matrix L is also integral and is called a *modular matrix of degree p* or, simply, a *modular matrix*. A square integral matrix M is called *unimodular* if M^{-1} exists and is integral. This is the case if and only if $|M| = \pm 1$. In particular, every modular matrix is unimodular, and the modular matrices form a subgroup Γ of the symplectic group called the *homogeneous modular group of degree p*, or simply, the *modular group*. It follows that the elements of the modular group are the integral solutions of (3).

The condition $C = 0$ determines a subgroup of the symplectic group whose elements, in view of (3), are determined by the equalities

$$AB' = BA', \qquad AD' = E.$$

It follows that we can take for A any real matrix with $|A| \neq 0$. Then $D' = A^{-1}$ and $A^{-1}B = (A^{-1}B)'$. It follows that we can take for $A^{-1}B = S$ any real symmetric matrix. Since $B = AS$, we find

$$L = \begin{pmatrix} A & AS \\ 0 & (A^{-1})' \end{pmatrix}. \tag{6}$$

The subgroup of the symplectic group formed by these matrices L will be denoted by Δ. The condition $A = E$ defines a subgroup Δ^*, which is a commutative normal subgroup of Δ. The factor group Δ/Δ^* can be represented by the elements L of Δ defined by the condition $S = 0$; these elements themselves form a subgroup of Δ. This formation of subgroups carries over directly to Γ. We see that A, A^{-1}, and AS in (6) must then be integral, so that A can be an arbitrary unimodular matrix and S, an arbitrary symmetric integral matrix.

We now define a collection of finitely many unimodular matrices and show that they generate the full modular group Γ. We define T_g $(g = 1, \ldots, p)$ as the matrix L in (6) with $A = E$ and S consisting of zeros except for the gth diagonal entry s_{gg} which is 1. Clearly, to obtain T_g^{-1} we need only replace S by $-S$, that is, $s_{gg} = 1$ by -1. Furthermore, we define U_g $(g = 1, \ldots, p)$

as the matrix in (6) with $S = 0$ and $A = (a_{kl})$, where $a_{kl} = 0$ $(k \neq l)$, $a_{kk} = 1$ $(k \neq g)$, $a_{gg} = -1$. Clearly, $U_g^{-1} = U_g$. Finally, for $p > 1$, we define the modular matrix U_{gh} $(g \neq h;\ g, h = 1, \ldots, p)$ by the conditions $S = 0$, $A = (a_{kl})$ with $a_{kk} = 1$ $(k = 1, \ldots, p)$, $a_{gh} = 1$ and $a_{kl} = 0$ $(k \neq l$, g or $l \neq k, h)$. This means that A arises from the unit matrix $E = (e_{kl})$ by replacing the element $e_{gh} = 0$ by 1. To obtain A^{-1} from A we need only replace $a_{gh} = 1$ by -1. There is a corresponding rule for the formation of U_{gh}^{-1}.

It is useful to specify the homogeneous linear transformations defined by the special modular matrices introduced above. Using the columns x_1 and x_2 defined in (4) we see that the transformation $x \to Lx$ splits into the transformations

$$x_1 \to Ax_1 + Bx_2, \qquad x_2 \to Cx_1 + Dx_2.$$

Let ξ_k and η_k $(k = 1, \ldots, p)$ denote the elements of x_1 and x_2. Putting for L successively the four matrices $T_g^{\pm 1}$, $(T_g^{\pm 1})'$, U_g, $U_g^{\pm 1}$, we obtain the four transformations

$$\xi_g \to \xi_g \pm \eta_g; \qquad \eta_g \to \eta_g \pm \xi_g; \qquad \xi_g \to -\xi_g, \quad \eta_g \to -\eta_g;$$
$$\xi_g \to \xi_g \pm \xi_h, \quad \eta_h \to \eta_h \mp \eta_g;$$

where, in each case, we have listed only the affected variables. Now let M be a matrix of order $2p$. To obtain LM from M in the four cases we must add the $(g + p)$th row to or subtract it from the gth row, add the gth row to or substract it from the $(g + p)$th row, multiply the gth row and the $(g + p)$th row by -1, add the hth row to or subtract it from the gth row and subtract the $(p + g)$th row from or add it to the $(h + p)$th row. *The individual transformations are said to be of type I, II, III, and IV.*

For $p = 1$, x is the column consisting of two numbers ξ, η, and these are transformed by transformations of type I and II into $\xi \pm \eta$, η and ξ, $\eta \pm \xi$. If ξ and η are nonzero real numbers, then

$$\min(|\xi + \eta| + |\eta|, |\xi - \eta| + |\eta|, |\xi| + |\eta + \xi|, |\xi| + |\eta - \xi|) < |\xi| + |\eta|.$$

If we call $|\xi| + |\eta|$ the *norm of the column* x, then it is possible to decrease the norm by one of these transformations provided that $\xi\eta \neq 0$. In particular, let ξ and η be integers. Then it follows by induction on the norm of x that, by repeated application of transformations of the first two types, it is possible to reduce one of the elements of the column to zero. If $\xi = 0$, then we take the pair $0, \eta$ into η, η by means of a transformation of type I, and η, η into $\eta, 0$ by means of a transformation of type II. It follows that we can always reduce the element η to 0.

Theorem 1: The homogeneous modular group of degree p is generated by the $p(p + 2)$ elements T_g, T_g', U_g, U_{gh}.

Proof: Let x be the first column of a given modular matrix M. For fixed $g = 1, \ldots, p$ a transformation of type I or II has the described effect on the two-entry column consisting of ξ_g, η_g and leaves the remaining elements fixed. Using this fact for each of the p values of g and bearing in mind the results above for $p = 1$, we see that it is possible, by repeated application of transformations of types I and II, to reduce the last p entries $\eta_1, \ldots, \eta_p$ in the first column x of L to 0. If $p > 1$, we effect further simplification of x by using transformations of type IV. The application of $U_{gh}^{\pm 1}$ and $U_{hg}^{\pm 1}$ means now that, for fixed distinct values of g and h in the sequence $1, \ldots, p$, the two-entry column consisting of ξ_g and ξ_h is subjected to transformations of the first two types, while the remaining elements of x stay fixed. In particular, choosing $g = 1$ and $h = 2, \ldots, p$ we can reduce $\xi_2, \ldots, \xi_p$ successively to 0. Finally, applying a transformation of type III for $g = 1$, we can carry ξ_1 into $-\xi_1$, and may therefore suppose $\xi_1 \geqslant 0$. Since T_g, U_g, U_{gh} all have determinant 1, the transformations applied thus far leave the determinant of L unchanged. If we remember that the elements $\xi_2, \ldots, \xi_p$ and $\eta_1, \ldots, \eta_p$ have all been reduced to 0 and that all the elements of a modular matrix are integers, then expansion of $|L|$ by the elements of the first column shows that ξ_1 is a factor of $|L| = \pm 1$, and so $\xi_1 = 1$. Thus x has been transformed into the first column of the $2p \times 2p$ unit matrix E_{2p}. By forming the alternating products with the remaining $2p - 1$ columns we find, by (5), that the $(1 + p)$th row of L coincides with the $(1 + p)$th row of E_{2p}.

Now we transform the $(1 + p)$th column of L. We denote this column by x and its elements by ξ_k, η_k $(k = 1, \ldots, p)$, and recall that now $\eta_1 = 1$. By repeated applications of transformations of type I with $g = 1$ we can reduce ξ_1 to 0. Using the reasoning in the preceding paragraph we see that $\xi_2, \ldots, \xi_p$ can be reduced to 0 by repeated applications of transformations of types I and II with $g = 2, \ldots, p$. Here we note that all of these transformations leave the first column and $(1 + p)$th row of L unchanged. This is also true when we repeatedly apply transformations of type IV with $g = 1$ and $h = 2, \ldots, p$ and reduce $\eta_2, \ldots, \eta_p$ to 0 while $\xi_1, \ldots, \xi_p$ retain the value 0. Now the $(1 + p)$th column also agrees with the $(1 + p)$th column of E_{2p}. By forming alternating products with the remaining columns we see, by (5), that the first row of L agrees with the first row of E_{2p}.

For $p = 1$ the proof is complete. For $p > 1$ we have the decomposition

$$(7) \qquad L = \begin{pmatrix} A & B \\ C & D \end{pmatrix}, \quad A = \begin{pmatrix} 1 & 0 \\ 0 & A_1 \end{pmatrix}, \quad B = \begin{pmatrix} 0 & 0 \\ 0 & B_1 \end{pmatrix},$$

$$C = \begin{pmatrix} 0 & 0 \\ 0 & C_1 \end{pmatrix}, \quad D = \begin{pmatrix} 1 & 0 \\ 0 & D_1 \end{pmatrix},$$

with $(p - 1) \times (p - 1)$ integral matrices A_1, B_1, C_1, D_1; in view of (3)

$$L_1 = \begin{pmatrix} A_1 & B_1 \\ C_1 & D_1 \end{pmatrix} \tag{8}$$

is seen to be a modular matrix of degree $(p - 1)$. Conversely, every modular matrix (8) yields a modular matrix of degree p of the special form (7) and, in particular, the matrices T_g, U_g, U_{gh}, formed with $(p - 1)$ instead of p, go over into the corresponding modular matrices of degree p. The assertion of our theorem follows by induction on p.

The proof of Theorem 1 also shows that every modular matrix has determinant 1. It is possible to give a brief proof of this fact for arbitrary symplectic matrices. The number $p(p + 2)$ of generators of the modular group found by us can be reduced by means of simple number-theoretical considerations. For our purpose, however, it is just these generators that are particularly useful by virtue of their simplicity.

5. *Canonical transformation*

To clarify the significance of the modular group for the period matrix, we must introduce the concept of *homology of closed curves* on the Riemann surface. In most cases this concept is defined in terms of combinatorial topology without reference to continuity or the fundamental group. In view of our extensive use of the concept of homotopy, we propose to make it the basis for the definition of homology. In this way the deeper relations and the connections with the theory of functions become clearer. We start from Sections 3 and 4 of Chapter 2 (Volume I).

When we introduced the fundamental group, it was essential that all the closed curves under consideration issued from a common point $\mathfrak{p}$ which remained fixed under all the continuous deformations. The only time when the choice of $\mathfrak{p}$ was superfluous was when we defined curves homotopic to zero. It is natural to ask for properties of closed curves on $\mathfrak{R}$ which are unaffected by arbitrary continuous deformations, including those which affect the initial point $\mathfrak{p}$ of a curve which is not homotopic to zero. One such property, called homology, can be defined using the fundamental group Ω of $\mathfrak{R}$. Thus, let C be an oriented, closed curve, $\mathfrak{q}$ a point on C, and $\mathfrak{p}$ the point used in defining Ω, which need not lie on C. We join $\mathfrak{p}$ to $\mathfrak{q}$ by means of a curve L such that LCL^{-1} is a closed curve issuing from $\mathfrak{p}$, and denote the homotopy class of LCL^{-1} by α. If M is another curve from $\mathfrak{p}$ to $\mathfrak{q}$ (Figure 37), then $Q = ML^{-1}$ is a closed curve with initial point $\mathfrak{p}$, and, in the sense of homotopy,

$$Q(LCL^{-1})Q^{-1} \backsim MCM^{-1}.$$

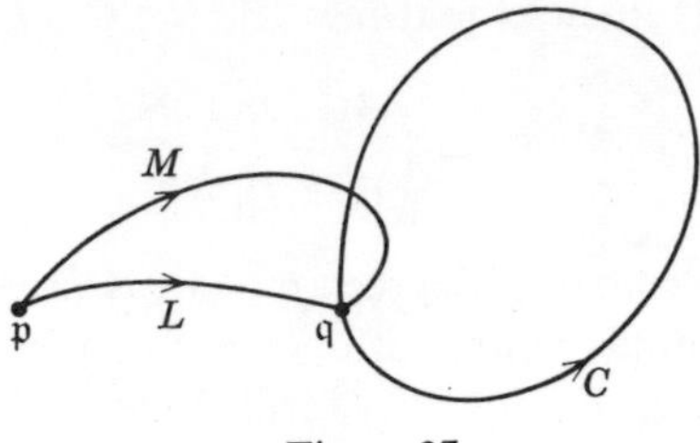

Figure 37

If β and γ are the homotopy classes of MCM^{-1} and Q, then

$$\beta = \gamma\alpha\gamma^{-1} = [\gamma, \alpha]\alpha.$$

It follows that α and β belong to the same coset of Ω relative to the commutator subgroup Λ, and so represent the same element of the commutative factor group Ω/Λ. We therefore associate with C the element in question, denote it by $\tilde{\alpha} = \tilde{\beta}$, and call it the *homology class* of C. By moving the point $\mathfrak{q}$ continuously on C to another point $\mathfrak{q}^*$, we establish the independence of the homology class of C from $\mathfrak{q}$ (Figure 38). This means that in defining C we can omit the initial point. Furthermore, if C_1 and C_2 are two curves with homology classes $\tilde{\alpha}_1$ and $\tilde{\alpha}_2$, then we join $\mathfrak{p}$ to C_1 and C_2 by means of two arcs L_1 and L_2, thus forming the closed curve $L_1C_1L_1^{-1}L_2C_2L_2^{-1}$ issuing from $\mathfrak{p}$. If we write the commutative homology group additively, then the homology class of the latter curve is $\tilde{\alpha}_1 + \tilde{\alpha}_2 = \tilde{\alpha}_2 + \tilde{\alpha}_1$. This gives meaning to $-C$ and $C_1 + C_2$. It is now clear that application of a continuous deformation to C does not affect the homology class of C even if the deformation in question leaves no point of C fixed; here C may, more generally, be the sum of finitely many closed curves. In such deformation we allow the replacement of two curves with a common point by a single curve, and, conversely, the separation of a curve with a double point into two curves (Figure 39); however, attention must be paid to orientation. The homology of two curves or two sums of curves C_1 and C_2 is denoted by $C_1 \backsim C_2$ and

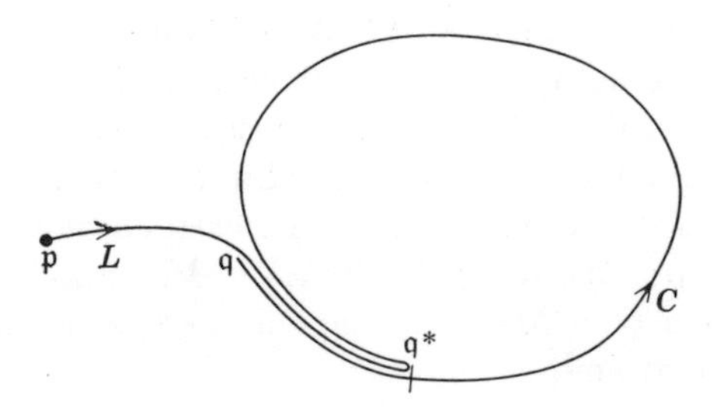

Figure 38

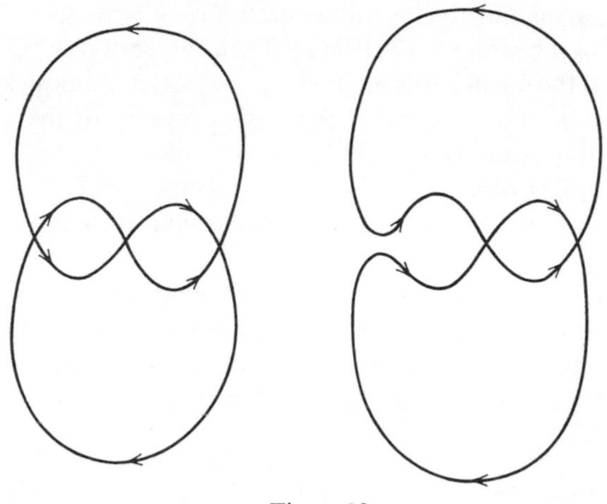

Figure 39

means that both belong to the same homology class. In particular, if C_2 is a point then C_1 is called *homologous to zero;* and C_1 can be contracted to a point by the now admissible deformations. More generally, $C_1 \backsim C_2$ if and only if, in the sense of the new definition, C_1 can be continuously deformed into C_2. In the sequel we shall integrate over the curves under consideration (it suffices to suppose them piecewise smooth). Cauchy's theorem implies that for every abelian integral of the first kind, the period $w(C)$ depends on the homology class of C alone.

We saw in Section 4 of Chapter 2 (Volume I) that the $2p$ curves $C_1, \ldots, C_{2p}$ of a canonical dissection of the Riemann surface $\mathfrak{R}$ yield a basis for the factor group Ω/Λ, which is an abelian group with $2p$ independent generators. It follows that we can associate uniquely with every curve C on $\mathfrak{R}$ $2p$ integers $g_1, \ldots, g_{2p}$ such that

$$C \backsim g_1C_1 + \cdots + g_{2p}C_{2p}. \tag{1}$$

The independence of $C_1, \ldots, C_{2p}$ in the sense of homology follows once more from the proof of

Theorem 1: A closed curve C is homologous to zero if and only if for every integral of the first kind w, the period $w(C)$ is equal to 0.

Proof: Without restriction of generality we may suppose the point $\mathfrak{p}$ on C. Then C is homotopic to a product of factors $C_k^{\pm 1}$ $(k = 1, \ldots, 2p)$. This yields the relation (1), if we mean by $g_1, \ldots, g_{2p}$ the corresponding exponent sums. It follows that

$$w(C) = g_1w(C_1) + \cdots + g_{2p}w(C_{2p}).$$

We now use, in particular, the $2p$ differentials dw_k $(k = 1, \ldots, 2p)$ constructed in the proof of Theorem 5, Section 2. Then the period $w_k(C)$ has the real part g_k, so that all the coefficients g_k $(k = 1, \ldots, 2p)$ are uniquely determined by the values $w(C)$, and, in view of the basis property of the dw_k, are all 0 if $w(C)$ is invariably equal to 0. The assertion follows.

Theorem 1 implies directly that $C_1 \backsim C_2$ if and only if $w(C_1) = w(C_2)$ for all w. It follows that the periods $w(C)$ determine the homology class of C uniquely.

Now, either a simple closed curve C on $\mathfrak{R}$ separates $\mathfrak{R}$ into two disjoint parts, or it does not do so. In the first case we say that C *bounds*. In the second case we can find another simple closed curve D such that C and D form a pair of conjugate crosscuts on $\mathfrak{R}$ and cross each other at a preassigned point of C.

Theorem 2: A simple closed curve C bounds if and only if it is homologous to zero.

Proof: Let C be the boundary of a domain $\mathfrak{B}$ on $\mathfrak{R}$. By making use of a dissection of $\mathfrak{R}$ into schlicht sheets over the complex sphere, we can put $\mathfrak{B}$ together out of finitely many nonoverlapping schlicht domains bounded by simple, closed curves $L_1, \ldots, L_n$. Since in the interior of $\mathfrak{B}$ any two abutting portions of the boundary have opposite orientations, we have

$$C \backsim L_1 + \cdots + L_n.$$

On the other hand, each of the curves L_k $(k = 1, \ldots, n)$ is homotopic to zero and, therefore, certainly homologous to zero. This implies that C is homologous to zero. If C does not bound, then we choose D so that C and D, and therefore also D^{-1} and C, are conjugate crosscuts. By Theorem 4, Section 2, there exists a differential of the first kind, dw, for which the period $w(C)$ has the real part $1 \neq 0$. By Theorem 1 this implies the remainder of our assertion.

The p pairs A_k, B_k $(k = 1, \ldots, p)$ of conjugate crosscuts of a canonical dissection of $\mathfrak{R}$ are characterized by the fact that they have in common the single point $\mathfrak{p}$ and that the dissected Riemann surface bounded by the curve

$$C = T_1 T_2 \cdots T_p, \qquad T_k = A_k B_k A_k^{-1} B_k^{-1} \qquad (k = 1, \ldots, p)$$

is simply connected. It is natural to ask if it is always possible to extend a given pair A_1, B_1 of conjugate crosscuts to a canonical dissection by the addition of $p - 1$ suitable pairs, and if this can be done in the more general case of q given pairs of conjugate crosscuts, $1 \leqslant q < p$. In this connection the investigation carried out for the case $p = 1$ in Section 10 of Chapter 1 (Volume I) is relevant.

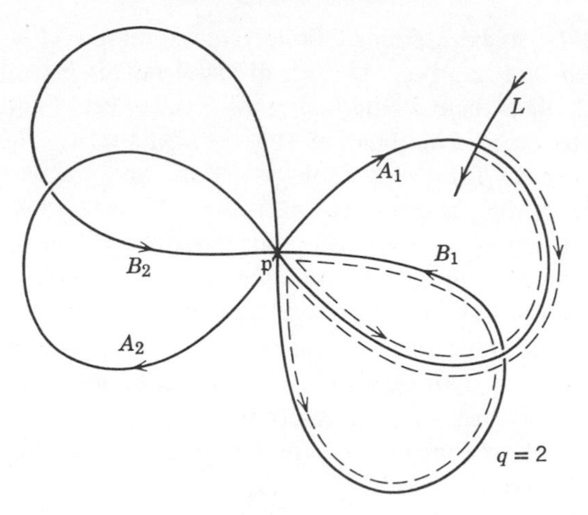

Figure 40

Theorem 3: Let A_k, B_k $(k = 1, \ldots, q)$ be q pairs of conjugate crosscuts with just one common point from which the $4q$ curves A_k, B_k^{-1}, A_k^{-1}, B_k issue in cyclic order, and let the dissected Riemann surface $\mathfrak{R}_q$ be bounded by the curve

$$C_q = T_1 T_2 \cdots T_q, \qquad T_k = A_k B_k A_k^{-1} B_k^{-1} \qquad (k = 1, \ldots, q)$$

be connected. Then $q \leqslant p$. Also, in the case $q = p$ the dissection is canonical, and in the case $q < p$ it can be extended to such by a suitable choice of $p - q$ additional pairs A_k, B_k $(k = q + 1, \ldots, p)$.

Proof: Using Theorem 4, Section 2, we construct $2q$ integrals of the first kind whose period matrix relative to the $2q$ curves A_k, B_k $(k = 1, \ldots, q)$ has the unit matrix E_{2q} as its real part. The argument in the proof of Theorem 5 in Section 2 implies that at least q of these abelian integrals are linearly independent over the complex numbers, so that, $q \leqslant p$. Now let L be a closed curve on $\mathfrak{R}$. We can continuously deform L in such a way that it has finitely many points in common with the curves A_k, B_k and cuts them at these points. At each such point of intersection, it is possible to get from one edge of the corresponding crosscut to the other by going around C_q (Figure 40). The additional path is invariably homologous to one of the curves $A_k^{\pm 1}$, $B_k^{\pm 1}$. Hence

$$(2) \qquad L \backsim L^* + \sum_{k=1}^{q} (g_k A_k + h_k B_k),$$

where g_k and h_k are integers, and L^* is a closed curve on the dissected surface $\mathfrak{R}_q$ which, therefore, does not meet the boundary C_q. After resolution of the double points in the manner explained in connection with Figure 39 we may suppose L^* to be a sum of finitely many simple closed curves on $\mathfrak{R}_q$ which may cut one another. If each of these curves bounded on $\mathfrak{R}$, then, by Theorem 2, all of them would be homologous to zero, so that L^* would be homologous to zero. But then, in view of (2), the A_k, B_k $(k = 1, \ldots, q)$ would form a basis for the homology group, and we would have $q = p$. If $q < p$, then among the distinct parts of L^*, there must be a curve A^* which does not bound on $\mathfrak{R}$ and does not meet C_q. It is possible to construct a crosscut B^* conjugate to A^* and, like A^*, not meeting C_q. Finally, we can shift the pair A^*B^* so as to move their common point to $\mathfrak{p}$. In this way we obtain an additional pair of curves A_{q+1}, B_{q+1}, thereby satisfying the assumption with $q + 1$ instead of q. The required $p - q$ pairs A_k, B_k $(k = q + 1, \ldots, p)$ are obtained by induction.

It remains to show that in the case $q = p$ the surface $\mathfrak{R}_p$ bounded by C_p is simply connected. To this end we take a second copy of $\mathfrak{R}_p$ reverse the orientation in it, and join the two surfaces edge to equal edge along the $4p$ edges of the cuts. One way of looking at this process is to think of $\mathfrak{R}_p$ as having a back which is added to the front. In this way we obtain a closed oriented surface $\mathfrak{R}_0$ which is not a Riemann region in the sense in which we have used the term so far. Nevertheless, the considerations of Section 2 of Chapter 2 (Volume I) carry over to $\mathfrak{R}_0$, so that $\mathfrak{R}_0$ is homeomorphic to a sphere, or to a sphere with handles. If $\mathfrak{R}_0$ were not a sphere, it would have positive genus and it would contain a closed curve M not homologous to zero on $\mathfrak{R}_0$. The boundary curve C_p of $\mathfrak{R}_0$ divides M into subarcs lying entirely on the front or back of $\mathfrak{R}_p$. This means that there is a simple, closed curve on $\mathfrak{R}_p$ which is not homologous to zero on $\mathfrak{R}_0$ and so is not homologous to zero on $\mathfrak{R}_p$. By Theorem 2, this curve does not bound on $\mathfrak{R}_p$. In view of the considerations in the preceding paragraph we have arrived at a contradiction. It follows that $\mathfrak{R}_0$ is homeomorphic to a sphere, and, since C_p is a simple, closed curve on the sphere $\mathfrak{R}_0$ which divides it into two simply connected regions, $\mathfrak{R}_p$ is simply connected. As a modification of this argument we could, instead of adding a back to $\mathfrak{R}_p$, attach a cover made of a schlicht polygon with $4p$ sides to the edge C_p of $\mathfrak{R}_p$. The procedure actually used has perhaps greater intuitive appeal.

In Section 9 of Chapter 2 (Volume I) we defined the *index* (A, B) of two oriented curves A and B. As part of the definition we assumed that A and B had only finitely many points in common. Now we suppose A and B closed. The points at which B cuts A from right to left are assigned the value $+1$, and the points at which B cuts A from left to right are assigned the value -1. The sum of all the ± 1 yields the index. Clearly, $(B, A) = -(A, B)$, $(-A, B) = -(A, B)$ and, furthermore,

$$(A_1 + A_2, B) = (A_1, B) + (A_2, B), \; (A, B_1 + B_2) = (A, B_1) + (A, B_2),$$

provided that the assumption on the finiteness of the number of points of intersection holds. Under this assumption the index is unchanged when A and B are continuously deformed; this shows that it depends only on the homology classes of A and B. It follows that the symbol (A, B) remains meaningful if we dispense with the assumption concerning the finiteness of the number of points of intersection, for this assumption can be satisfied by means of arbitrarily small deformations of the curves. In particular, we can form (A, A) and obtain for it the value 0. If $C_1, \ldots, C_m$ are closed curves on $\mathfrak{R}$, then by their *intersection matrix* we mean the skew-symmetric matrix with the m^2 integer entries (C_k, C_l) $(k, l = 1, \ldots, m)$. In particular, let $C_1, \ldots, C_{2p}$ be the curves $A_1, \ldots, A_p$ and $B_1, \ldots, B_p$ of a canonical dissection in this order. Since, in this case, C_k and C_l with $k < l$ cut each other at the common point $\mathfrak{p}$ only if $l = k + p$, we have

$$(3) \quad (C_k, C_l) = 0 \quad (l - k \neq \pm p), \qquad (C_k, C_l) = \pm 1 \quad (l - k = \pm p).$$

It follows that the intersection matrix is just the matrix J defined in (4), Section 3.

Theorem 4: Let $C_k = A_k$, $C_{k+p} = B_k$ $(k = 1, \ldots, p)$ and $C_k^* = A_k^*$, $C_{k+p}^* = B_k^*$ $(k = 1, \ldots, p)$ be the pairs of crosscuts of two canonical dissections of $\mathfrak{R}$. In the sense of homology we have the $2p$ relations

$$(4) \qquad C_k^* \backsim \sum_{l=1}^{2p} m_{kl} C_l \qquad (k = 1, \ldots, 2p)$$

with $4p^2$ uniquely determined integral coefficients m_{kl} which form a modular matrix $(m_{kl}) = M$.

Proof: Since $C_1, \ldots, C_{2p}$ yields a basis for the homology group of $\mathfrak{R}$, (4) holds for certain definite integers m_{kl} $(k, l = 1, \ldots, 2p)$. If m_l denotes the lth column of the matrix M, then, in view of (3) and (4), the index (C_k^*, C_l^*) is just the alternating product of m_k and m_l. Since, in addition,

$$(C_k^*, C_l^*) = 0 \quad (l - k \neq \pm p), \qquad (C_k^*, C_l^*) = \pm 1 \quad (l - k = \pm p),$$

M satisfies the very conditions (5) in Section 4 required of a modular matrix.

We now come to the converse of Theorem 4.

Theorem 5: Let $C_k = A_k$, $C_{k+p} = B_k$ $(k = 1, \ldots, p)$ be the pairs of crosscuts of a canonical dissection of $\mathfrak{R}$ and let $M = (m_{kl})$ be a modular matrix. Then there exists an additional canonical dissection of $\mathfrak{R}$ whose pairs of crosscuts $C_k^* = A_k^*$, $C_{k+p}^* = B_k^*$ $(k = 1, \ldots, p)$ satisfy the conditions (4).

Proof: In view of Theorem 1 in the preceding section, we need only establish our assertion for the cases when $M^{\pm 1}$ is one of the generators

T_g, T'_g, U_g, U_{gh} of the modular group. In these cases, we shall give the required second canonical dissection directly. First, we suppose the p pairs of conjugate crosscuts A_k, B_k detached by a continuous deformation from the point $\mathfrak{p}$ common to all of them so that the individual pairs are separated from each other, and denote the common point of A_k and B_k by $\mathfrak{p}_k$ $(k = 1, \ldots, p)$.

The case $M = U_g = U_g^{-1}$ is easiest to settle, for U_g was the matrix of the substitution $\xi_g \to -\xi_g$, $\eta_g \to -\eta_g$ with the remaining variables unchanged. If we put, accordingly,

$$A_k^* = A_k, \qquad B_k^* = B_k \qquad (k \neq g), \qquad A_g^* = A_g^{-1}, \qquad B_g^* = B_g^{-1},$$

then, clearly, A_g^*, B_g^* is also a pair of crosscuts separated from the remaining $p - 1$ pairs. By suitably bringing the points $\mathfrak{p}_k$ together at $\mathfrak{p}$, we obtain, in this case, the required result.

For $M = T_g^{\pm 1}$ we must consider the two substitutions: $\xi_g \to \xi_g + \eta_g$ and $\xi_g \to \xi_g - \eta_g$. The closed curve $A_g B_g$ has a double point at $\mathfrak{p}_g$ where, however, it does not cut itself. A small deformation turns $A_g B_g$ into a simple, closed curve A_g^*. By means of another small deformation, we take B_g into a neighboring simple, closed curve B_g^* which cuts A_g^* at $\mathfrak{p}_g$, and does not meet it otherwise (Figure 41). Then A_g^*, B_g^* form a pair of conjugate crosscuts, and

$$A_g^* \backsim A_g + B_g, \qquad B_g^* \backsim B_g.$$

Similarly, we can derive another pair of conjugate crosscuts A_g^*, B_g^* from $A_g B_g^{-1}$, B_g (Figure 42). We again put $A_k^* = A_k$, $B_k^* = B_k$ $(k \neq g)$, bring all the points $\mathfrak{p}_k$ $(k = 1, \ldots, p)$ together at $\mathfrak{p}$, and in this way settle the case under consideration. The case $M' = T_g^{\pm 1}$ is reduced to the case just considered by replacing A_k, B_k for $k = 1, \ldots, p$ by $B_k^{\pm 1}$, $A_k^{\mp 1}$. To this substitution there corresponds for the homology classes the modular matrix $J^{\pm 1}$, and we have $T'_g = J T_g^{-1} J^{-1}$.

Finally, for $M = U_{gh}$ we have to consider the substitution $\xi_g \to \xi_g + \xi_h$, $\eta_g \to \eta_g - \eta_h$ $(g \neq h)$. Here we join $\mathfrak{p}_g$ and $\mathfrak{p}_h$ by means of a simple curve L

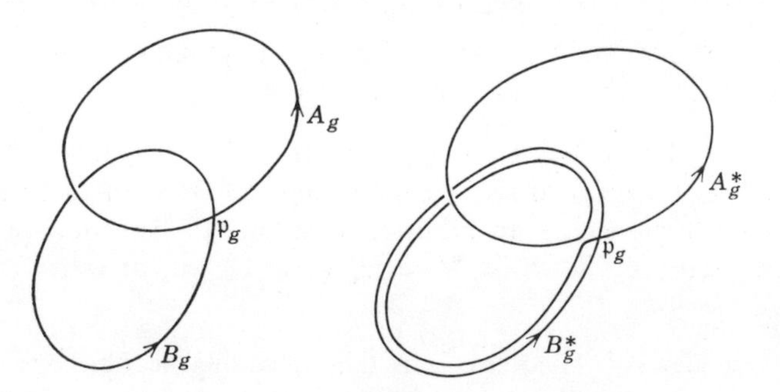

Figure 41

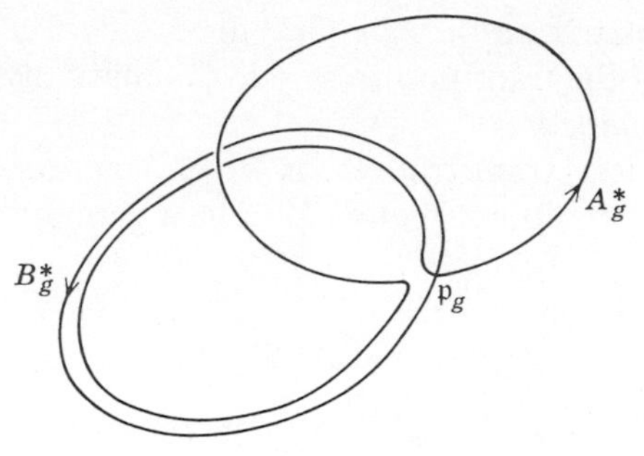

Figure 42

which does not otherwise meet the given crosscuts, issues from $\mathfrak{p}_g$ in the third quadrant, and terminates at $\mathfrak{p}_h$ in the fourth quadrant (Figure 43). The pairs of crosscuts A_g^*, B_g^* and A_h^*, B_h^* arise from the curves $A_gLA_hL^{-1}$, B_g and A_h, $B_hL^{-1}B_gL$ as a result of the small deformation in the figure, and we have, in accordance with requirements,

$$A_g^* \backsim A_g + A_h, \qquad B_g^* \backsim B_g, \qquad A_h^* \backsim A_h, \qquad B_h^* \backsim B_h - B_g.$$

Since $U_{hg} = U'_{gh}$ and $U_{gh}^{-1} = J^{-1}U'_{gh}J$, the remaining case $M = U_{gh}^{-1}$ is thereby also settled and the proof is complete.

Theorems 4 and 5 settle completely the question of the possible connections between the homology classes of curves belonging to two canonical dissections. The corresponding problem for the homotopy classes is far more difficult for $p > 1$ and is essentially unsolved. Now we turn to the problem mentioned at the beginning of Section 4 concerning the dependence of the

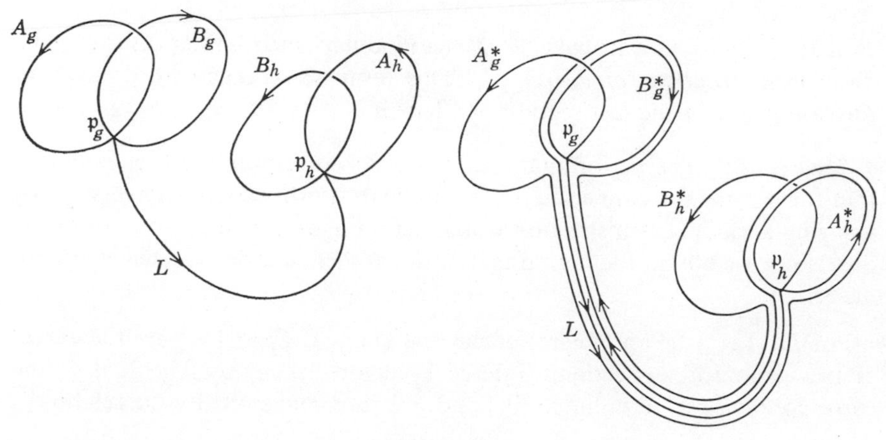

Figure 43

normalized period matrix on the canonical dissection of $\mathfrak{R}$. For the sake of brevity, we shall call the transition from one canonical dissection to another a *canonical transformation.*

Let Z be a complex symmetric matrix of order p, and Y its imaginary part. To begin with we do not require Z to be a period matrix. If

$$T = \begin{pmatrix} A & B \\ C & D \end{pmatrix}$$

is symplectic, (2) in Section 4 implies the relation

$$(5) \qquad (C\bar{Z} + D)'(AZ + B) - (A\bar{Z} + B)'(CZ + D) = Z - \bar{Z} = 2iY.$$

Now let $Y > 0$ and let the complex column s be a solution of

$$(CZ + D)s = 0.$$

By (5) we then have $\bar{s}'Ys = 0$, and so $s = 0$. Hence the determinant $|CZ + D|$ is different from 0 and it makes sense to form the matrix

$$(6) \qquad (AZ + B)(CZ + D)^{-1} = Z^*.$$

We have

$$(CZ + D)'(Z^* - (Z^*)')(CZ + D) = (CZ + D)'(AZ + B)$$
$$-(AZ + B)'(CZ + D) = Z - Z' = 0.$$

It follows that the $p \times p$ complex matrix Z^* is symmetric and for its imaginary part Y^* we have, by (5), the formula

$$(C\bar{Z} + D)'Y^*(CZ + D) = \frac{1}{2i}(C\bar{Z} + D)'(Z^* - (\bar{Z}^*)')(CZ + D) = Y;$$

so that $Y^* > 0$. In particular, if T is a modular matrix, then (6) is called a *modular substitution* (*of degree p*). This term was already introduced for the case $p = 1$ in the last section of Chapter 3.

Theorem 6: The period matrices Z and Z^* corresponding to a canonical transformation are connected by a modular substitution. Conversely, by applying a modular substitution to the period matrix Z of a given canonical dissection, we obtain the period matrix of another suitably selected canonical dissection.

Proof: Let w be an integral of the first kind. If C_k and C_k^* are the curves of two canonical dissections ordered as before, then, by Theorem 4, the corresponding period columns $[w]$ and $[w]^*$ are connected by the relation

$$[w]^* = M[w]$$

with M a modular matrix uniquely determined by the canonical transformation. If we choose a fixed complex basis $w_1, \ldots, w_p$ for the integrals of the first kind, then we obtain for the corresponding period matrices

$$Q = \begin{pmatrix} F \\ G \end{pmatrix}, \qquad Q^* = \begin{pmatrix} F^* \\ G^* \end{pmatrix}$$

the formula

$$Q^* = MQ. \tag{7}$$

If we put

$$M = \begin{pmatrix} A & B \\ C & D \end{pmatrix}, \qquad GF^{-1} = Z, \qquad G^*(F^*)^{-1} = Z^*,$$

then we have

$$F^* = AF + BG, \qquad G^* = CF + DG,$$
$$Z^* = (CF + DG)(AF + BG)^{-1} = (DZ + C)(BZ + A)^{-1}.$$

By (2), Section 4, if M is a modular matrix, then so is

$$N = \begin{pmatrix} D & C \\ B & A \end{pmatrix}.$$

Hence Z^* is obtained from Z by a modular substitution. Conversely, given a modular matrix N and a canonical dissection with corresponding normalized period matrix

$$Q = \begin{pmatrix} E \\ Z \end{pmatrix},$$

we can, by Theorem 5, find another canonical dissection such that the corresponding period matrix Q^*, formed for the same integrals, is given by (7). Normalization of Q^* again yields

$$Z^* = (DZ + C)(BZ + A)^{-1},$$

and the proof is complete.

If we view the entries z_{kl} $(1 \leqslant k \leqslant l \leqslant p)$ of the complex symmetric matrix $(z_{kl}) = Z = X + iY$ as independent variables then their number is $p(p + 1)/2 = m$. The condition $Y > 0$ defines in the space of m complex variables a region $\mathfrak{H}$ which we shall call the *generalized upper half plane* or the *upper half plane of degree* p; this is so because for $p = 1$ the region in question is just the upper half plane. As shown earlier, the modular substitutions and, more generally, the symplectic substitutions (6) map the region $\mathfrak{H}$ onto itself; and the elements of Z^* are rational functions of the m variables z_{kl} which are everywhere regular on $\mathfrak{H}$. It follows that the modular substitutions yield a representation of the homogeneous modular group by means of a group Γ_0 of regular mappings of the generalized upper half plane onto itself. Here we note that for every symplectic matrix T, $-T$ defines the same substitution (6). On the other hand, if (6) is the identity mapping of $\mathfrak{H}$ onto itself, then the equation

$$ZCZ + ZD = AZ + B$$

holds identically in all the m variables $z_{kl} = z_{lk}$. By equating coefficients we see that $C = B = 0$ and $A = D = \lambda E$ with scalar λ, so that, in view of $AD' - BC' = E$, $\lambda = \pm 1$. It follows that Γ_0 arises as a factor group from Γ, if we identify all modular matrices T with $-T$. We call Γ_0 the *inhomogeneous modular group of degree p* or, simply, the *modular group*. For $p = 1$ this group turned up in Section 9 of Chapter 3. Its detailed investigation will be undertaken in Chapter 6 (Volume III). In that chapter we introduce, in particular, the functions of m complex variables z_{kl} which are meromorphic in $\mathfrak{H}$ and are invariant under all the modular substitutions of degree p; as such they are called *modular functions of degree p*. In particular, the case $p = 1$ covers the *elliptic modular functions*, which therefore depend on just one complex variable z.

By Theorem 1 of the preceding section, the generators of the homogeneous elliptic modular group are the three matrices

$$T = \begin{pmatrix} 1 & 1 \\ 0 & 1 \end{pmatrix}, \qquad T' = \begin{pmatrix} 1 & 0 \\ 1 & 1 \end{pmatrix}, \qquad U = \begin{pmatrix} -1 & 0 \\ 0 & -1 \end{pmatrix}.$$

If we put, in addition,

$$V = T(T')^{-1} = \begin{pmatrix} 1 & 1 \\ 0 & 1 \end{pmatrix}\begin{pmatrix} 1 & 0 \\ -1 & 1 \end{pmatrix} = \begin{pmatrix} 0 & 1 \\ -1 & 1 \end{pmatrix}$$

and note the equalities

$$V^2T' = VT = \begin{pmatrix} 0 & 1 \\ -1 & 0 \end{pmatrix} = J, \qquad V^3 = \begin{pmatrix} -1 & 1 \\ -1 & 0 \end{pmatrix}\begin{pmatrix} 0 & 1 \\ -1 & 1 \end{pmatrix} = U,$$

then we can express the generators T, T', U in terms of J and V in the form

$$T = V^{-1}J, \qquad T' = V^{-2}J, \qquad U = J^2 = V^3,$$

which shows that the homogeneous elliptic modular group is generated by J and V. The corresponding modular substitutions are

$$z^* = \frac{-1}{z}, \qquad z^* = \frac{1}{1 - z}.$$

If we denote these substitutions by ι and κ and note that U corresponds to the identity ε, then we see that the inhomogeneous elliptic modular group has the two generators ι and κ which satisfy the relations $\iota^2 = \varepsilon$, $\kappa^3 = \varepsilon$. It can be shown that all other relations connecting ι and κ follow from these two defining relations, so that a product of factors of the form $\iota\kappa^{\pm 1}$ is never ε. This yields a unique description of the inhomogeneous elliptic modular group as an abstract group.

Theorem 3 in Section 9 of Chapter 1 (Volume I) and the subsequent remark showed that the domain $\mathfrak{B}$ in the upper half plane $\mathfrak{H}$ of the complex variable $z = x + iy$ defined by $-\frac{1}{2} \leqslant y \leqslant \frac{1}{2}$, $x^2 + y^2 \geqslant 1$ is a fundamental region of the elliptic modular group in the sense that the images of $\mathfrak{B}$ under all the different modular substitutions yield a simple covering of $\mathfrak{H}$ without gaps. Furthermore, in Section 9 of Chapter 3 there appeared an elliptic modular function, namely, the function of the period ratio $\tau = \omega_2/\omega_1$ denoted there by $j(\tau)$. Its essential property is that it is regular on $\mathfrak{B}$ and takes on in $\mathfrak{B}$ every complex value exactly once, provided that we leave out the right half of the boundary of $\mathfrak{B}$ with the exception of i, as shown in Figure 22 of Volume I. On the other hand, every elliptic function field is determined by the value $j(\tau)$ in a one-to-one manner. Later we shall see in a more general context that, with an additional assumption concerning its behavior for $z \to \infty$, every elliptic modular function is a rational function of $j(z)$ with constant coefficients. This is why, when there is no danger of confusion, $j(z)$ is referred to simply as the *elliptic modular function.*

6. *The theorem of Riemann and Roch*

In this and in the next section, the value $p = 0$ is again admissible. When investigating the functions in the field K, we must distinguish between a property which holds regardless of the choice of the independent variable x and a property which is meaningful only when the Riemann surface $\mathfrak{R}$ is held fixed. Examples of properties of the latter type are statements concerning the branches of a function. On the other hand, the number of poles of a given function is independent of the choice of $\mathfrak{R}$. The problems which we are about to treat belong in the latter category. It follows that, insofar as our present purposes are concerned, we could replace the concept of a point of a fixed Riemann surface with an abstract concept which would remain invariant under identification of all Riemann surfaces which can be mapped conformally onto $\mathfrak{R}$. However, confusion could arise in the case of conformal mappings of $\mathfrak{R}$ onto itself. Also, when it comes to proofs, we shall find it convenient for the formulas to contain definite local parameters. That is why we shall continue to make use of a fixed Riemann surface $\mathfrak{R}$.

Let $\mathfrak{p}_1, \ldots, \mathfrak{p}_g$ be not necessarily distinct points on $\mathfrak{R}$. We introduce the symbol

$$\mathfrak{d} = \mathfrak{p}_1 \mathfrak{p}_2 \cdots \mathfrak{p}_g$$

and call it an *integral divisor*. An integral divisor consisting of one factor, that is, a single point $\mathfrak{p}$, will also be called a *prime divisor*. In this definition of $\mathfrak{d}$, the order of the factors is irrelevant. By a *divisor* $\mathfrak{d}$ we mean a formal quotient

$$\mathfrak{d} = \frac{\mathfrak{d}_1}{\mathfrak{d}_2} = \frac{\mathfrak{p}_1 \mathfrak{p}_2 \cdots \mathfrak{p}_g}{\mathfrak{q}_1 \mathfrak{q}_2 \cdots \mathfrak{q}_h}, \qquad \mathfrak{d}_1 = \mathfrak{p}_1 \mathfrak{p}_2 \cdots \mathfrak{p}_g, \qquad \mathfrak{d}_2 = \mathfrak{q}_1 \mathfrak{q}_2 \cdots \mathfrak{q}_h$$

of integral divisors $\mathfrak{d}_1$ and $\mathfrak{d}_2$. If we carry over the usual rules of reduction and multiplication of fractions to divisors, then the latter form an infinite commutative multiplicative group whose independent generators are the prime divisors. As unit element $\mathfrak{e}$ we take the divisor for which, in reduced form, $g = h = 0$. By the *degree* of $\mathfrak{d}$ we mean the integer $g - h$ which is invariant under reduction. It is clear that the divisors of degree 0 form a group in their own right.

The functions f meromorphic on $\mathfrak{R}$ form the algebraic function field K. By the *order* of f we mean the number m of poles of f on $\mathfrak{R}$ with each pole counted according to its multiplicity. We use the notation

$$\mathfrak{O}(f) = m.$$

If $m = 0$, then f is regular on all of $\mathfrak{R}$, hence, constant. Conversely, for every constant f, m has the value 0.

Theorem 1: With multiplicities taken into consideration, every meromorphic function on $\mathfrak{R}$ of positive order m takes on every complex value exactly m times.

Proof: Let f be the given function and c a constant. Then $\mathfrak{O}(f - c) = m$ and, on the other hand, $d \log (f - c)$ is, by Theorem 5 in Section 1, a differential of the third kind whose sum of residues is, by Theorem 1, Section 1, equal to 0. This residue sum is just the difference between the number of zeros and the number of poles for the function $f - c$. Our assertion follows.

This theorem is also a consequence of Theorem 3, Section 14 of Chapter 1, and Theorem 1, Section 6 of Chapter 2 in conjunction with the results of Section 7 of Chapter 2.

Let f be a function in K of positive order m with zeros $\mathfrak{p}_1, \ldots, \mathfrak{p}_m$ and poles $\mathfrak{q}_1, \ldots, \mathfrak{q}_m$ counting multiplicities. Then the divisor

$$\mathfrak{d} = \frac{\mathfrak{p}_1 \cdots \mathfrak{p}_m}{\mathfrak{q}_1 \cdots \mathfrak{q}_m} \tag{1}$$

is uniquely determined by f, and we put

$$\mathfrak{d} = (f). \tag{2}$$

If f is a nonzero constant, then we put $(f) = \mathfrak{e}$. No divisor corresponds to the constant 0. If f and g are two nonzero functions in K, then, clearly,

$$(fg) = (f)(g), \qquad (fg^{-1}) = (f)(g)^{-1},$$

and $(f) = (g)$ implies not the equality $f = g$, but the relation $f = cg$, with constant $c \neq 0$. We call (f) the *divisor associated with* f and note that (f) determines f to within a nonzero constant. If a given divisor $\mathfrak{d}$ is associated with a function f in K, then we call it a *principal divisor*. In the next section we shall establish necessary and sufficient conditions for a given divisor $\mathfrak{d}$ to be a principal divisor. A necessary condition follows from (1); it is that $\mathfrak{d}$ must be of degree 0.

To begin with we consider a related issue. Let $\mathfrak{d}$ be a divisor and f a function in K different from 0. If $\mathfrak{d}^{-1}(f)$ is an integral divisor, then we call $\mathfrak{d}$ a *factor* of f, and write $\mathfrak{d} \mid f$. Let us put, for the sake of brevity,

$$\mathfrak{d} = \frac{\mathfrak{p}_1 \cdots \mathfrak{p}_g}{\mathfrak{q}_1 \cdots \mathfrak{q}_h}.$$

The formula $\mathfrak{d} \mid f$ means that, counting multiplicities, the points $\mathfrak{p}_1, \ldots, \mathfrak{p}_g$ are among the zeros of f, and the poles of f are among the points $\mathfrak{q}_1, \ldots, \mathfrak{q}_h$. If we use this terminology we need no longer exclude the case when f is identically 0, as we now admit the convention $\mathfrak{d} \mid 0$ for all $\mathfrak{d}$. It is now clear that $\mathfrak{d} \mid f$ and $\mathfrak{d} \mid g$ imply $\mathfrak{d} \mid \lambda f + \mu g$ for all complex constants λ, μ. It follows that the functions f in K for which $\mathfrak{d} \mid f$ form a vector space over the field of complex numbers. Our immediate aim is to determine its dimension. To treat this question we must introduce for the abelian differentials the concepts of *associated divisors* and *divisibility*. It is clear how this should be done. Specifically, if dw is a differential which is not identically zero with zeros at $\mathfrak{p}_1, \ldots, \mathfrak{p}_g$ and poles at $\mathfrak{q}_1, \ldots, \mathfrak{q}_h$, then, in analogy to (1) and (2), we define the divisor

$$\frac{\mathfrak{p}_1 \cdots \mathfrak{p}_g}{\mathfrak{q}_1 \cdots \mathfrak{q}_h} = (dw).$$

This divisor is integral if and only if dw is of the first kind. In analogy to $\mathfrak{d} \mid f$, $\mathfrak{d} \mid dw$ is to mean that dw is identically 0 or the divisor $\mathfrak{d}^{-1}(dw)$ is integral. This means that the zeros of dw include the prime divisors in the reduced numerator of $\mathfrak{d}$, and the poles of dw are included among the prime divisors in the reduced denominator of $\mathfrak{d}$. It is clear that the differentials defined by the condition $\mathfrak{d} \mid dw$ for a fixed $\mathfrak{d}$ likewise form a vector space. Furthermore, the equality $(dw_1) = (dw_2)$ for two differentials dw_1 and dw_2

implies that their ratio dw_1/dw_2 is a function in K without zeros or poles and, therefore, a constant c, so that $dw_1 = c\,dw_2$. An answer to the question stated above is given by the Riemann–Roch theorem which we state first for the special case when $\mathfrak{d}^{-1}$ is an integral divisor. For this divisor we shall again use the symbol $\mathfrak{d}$.

Theorem 2: Let $\mathfrak{d}$ be an integral divisor of degree m. Then the dimensions of the two complex vector spaces defined by $\mathfrak{d}^{-1} \mid f$ and $\mathfrak{d} \mid dw$ are finite. If a and b are the dimensions in question, then we have

$$a - b = m - p + 1.$$

Proof: The corresponding dimensions over the field of real numbers are $a_0 = 2a$ and $b_0 = 2b$. We must show that a_0 and b_0 are finite and satisfy the equation

$$a_0 - b_0 = 2m - 2p + 2. \tag{3}$$

The proof is in three steps.

I. If $\mathfrak{p}_1, \ldots, \mathfrak{p}_n$ are the distinct prime divisors which are factors of $\mathfrak{d}$, then we write

$$\mathfrak{d} = \mathfrak{p}_0^{l_1} \cdots \mathfrak{p}_n^{l_n}$$

with natural exponents $l_1, \ldots, l_n$ satisfying the condition

$$l_1 + \cdots + l_n = m. \tag{4}$$

Using the differentials $dE_l(\mathfrak{x}_1)$, $dF_l(\mathfrak{x}_1)$ of the second kind defined in Theorem 1, Section 2, and arbitrary real coefficients λ_{kl}, μ_{kl} ($l = 1, \ldots, l_k$; $k = 1, \ldots, n$), we form the sum

$$d\varphi = \sum_{k=1}^{n} \sum_{l=1}^{l_k} (\lambda_{kl}\, dF_l(\mathfrak{p}_k) + i\mu_{kl}\, dE_l(\mathfrak{p}_k)). \tag{5}$$

It is clear that (5) is a differential of the second or first kind which may have a pole at $\mathfrak{p}_k$ ($k = 1, \ldots, n$) of order not exceeding $l_k + 1$, and has no other poles. Also, the principal parts at the various poles are determined in a biunique manner by the complex numbers $\lambda_{kl} + i\mu_{kl}$. Let $\mathfrak{x}_0$ be a point on $\mathfrak{R}$ different from $\mathfrak{p}_1, \ldots, \mathfrak{p}_n$. If the integral φ on $\mathfrak{R}$ with lower limit $\mathfrak{x}_0$ is single-valued, then, for every complex constant $c = \lambda_0 + i\mu_0$, the function $f = \varphi + c$ is in K, and we have $f(\mathfrak{x}_0) = c$, $\mathfrak{d}^{-1} \mid f$. Conversely, if $\mathfrak{d}^{-1} \mid f$, then it is possible to determine the λ_{kl} and μ_{kl} in (5) uniquely so that $d\varphi$ and df have the same principal parts at all poles. If we put

$$f(\mathfrak{x}_0) = c, \qquad f - c - \varphi = \chi,$$

then χ is an integral of the first kind whose imaginary part is, by Theorem 1 in Section 2, single-valued and harmonic on $\mathfrak{R}$. It follows that $\chi = 0$. We see that the required functions f are given by $f = \varphi + \lambda_0 + i\mu_0$ with arbitrary real λ_0 and μ_0, provided that φ is an integral defined by (5) which is single-valued on $\mathfrak{R}$.

In the case $p = 0$, $\mathfrak{R}$ is simply connected, so that the requirement of single-valuedness of φ is automatically satisfied for every choice of the coefficients λ_{kl}, μ_{kl}. The number of these coefficients is $2(l_1 + \cdots + l_n)$, and, since we must count the two coefficients λ_0, μ_0, (4) implies that in the present case

$$a_0 = 2(l_1 + \cdots + l_n) + 2 = 2m + 2. \tag{6}$$

On the other hand, for $\mathfrak{d} \mid dw$ the differential dw must be of the first kind, and, since $p = 0$, identically zero. This implies that $b_0 = 0$. Now (3) is easily seen to hold for $p = 0$; so that, in the sequel, we suppose $p > 0$.

II. We start with a fixed canonical dissection of $\mathfrak{R}$ whose curves avoid the given points $\mathfrak{p}_1, \ldots, \mathfrak{p}_n$. The requirement of single-valuedness of φ on $\mathfrak{R}$ is expressed by the formula $[\varphi] = 0$. This requirement is now reformulated as follows. Let $dw_1, \ldots, dw_{2p}$ be the differentials of the first kind used in the proof of Theorem 5 in Section 2. Then the real part of the matrix

$$W = ([w_1] \cdots [w_{2p}])$$

is E_{2p}, and every differential dw of the first kind can be uniquely expressed as a linear combination

$$dw = v_1\, dw_1 + \cdots + v_{2p}\, dw_{2p} \tag{7}$$

with real coefficients $v_1, \ldots, v_{2p}$. Now let φ again be given by (5). The expression $\varphi\, dw$ is single-valued on the canonically dissected surface $\mathfrak{R}^*$ and, since none of the points $\mathfrak{p}_1, \ldots, \mathfrak{p}_n$ lies on the boundary of $\mathfrak{R}^*$, can be integrated over C. With $\psi = w$ formula (3) of Section 3 holds in the present case of an integral φ of the second kind and yields

$$\frac{1}{2\pi i} \int_C \varphi\, dw = \frac{1}{2\pi i} [\varphi]' J [w]. \tag{8}$$

On the other hand, by the residue theorem, the left-hand side of (8) is equal to the sum of the residues of $\varphi\, dw/d\zeta$ on $\mathfrak{R}^*$, where ζ is the appropriate local parameter. Since $dw/d\zeta$ is regular, and the multiple-valuedness of φ results only from the addition of constants, namely the periods, every such residue is independent of the canonical dissection used. If $\rho_k(\varphi\, dw)$ denotes the real part of the residue of $\varphi\, dw/d\zeta$ at the point $\mathfrak{p}_k$ $(k = 1, \ldots, n)$, then the value

$$\sum_{k=1}^{n} \rho_k(\varphi\, dw) = \rho(\varphi\, dw) = \rho \tag{9}$$

is equal to the real part of the residue sum under consideration. We denote by ν the column of coefficients $\nu_1, \ldots, \nu_{2p}$ in (7). Then

$$[w] = W\nu,$$

and, since $[\varphi]$ is real in view of Theorem 1 in Section 2 and definition (5), the real part of the alternating product of $[\varphi]$ and $[w]$ is equal to the alternating product of $[\varphi]$ and ν. Now (8) and (9) imply that φ is single-valued on $\mathfrak{R}$ if and only if $\rho = 0$ for all differentials dw of the first kind. In order to express this condition somewhat more clearly, we denote the $2m$ integrals $F_l(\mathfrak{p}_k)$, $iE_l(\mathfrak{p}_k)$ in definite order by $e_1, \ldots, e_{2m}$, and, correspondingly, the real coefficients λ_{kl}, μ_{kl} in (5) by $\lambda_1, \ldots, \lambda_{2m}$, with the latter arranged in a column λ. If R is the matrix with the real entries

$$r_{kl} = \rho(e_k\, dw_l) \qquad (k = 1, \ldots, 2m;\ \ l = 1, \ldots, 2p),$$

then the condition in question is

$$\lambda' R\nu = 0$$

for arbitrary real $\nu_1, \ldots, \nu_{2p}$; this means that

$$R'\lambda = 0. \tag{10}$$

If r is the rank of R, and so also of R', then equation (10) has exactly $2m - r$ linearly independent real solutions λ. Since, in addition, λ_0, μ_0 may be chosen at will, it follows that

$$a_0 = 2m - r + 2. \tag{11}$$

III. We now study the significance of the equation

$$R\nu = 0 \tag{12}$$

in the $2p$ real unknowns $\nu_1, \ldots, \nu_{2p}$. This equation is satisfied if and only if the differential dw of the first kind defined by (7) satisfies the conditions

$$\rho(e_s\, dw) = 0 \qquad (s = 1, \ldots, 2m). \tag{13}$$

Let

$$\frac{dw}{d\zeta} = c_0 + c_1\zeta + \cdots$$

be the power series at the point $\mathfrak{p}_k$ $(k = 1, \ldots, n)$. In particular, if, for a fixed k, we put for e_s the $2l_k$ integrals $F_l(\mathfrak{p}_k)$, $iE_l(\mathfrak{p}_k)$ $(l = 1, \ldots, l_k)$, then the corresponding conditions (13) are equivalent to

$$c_{l-1} = 0 \qquad (l = 1, \ldots, l_k)$$

and the latter state that the differential dw vanishes at the point $\mathfrak{p}_k$ at least to order l_k. Therefore (12) states that $\mathfrak{d} \mid dw$. But (12) has exactly $2p - r$ linearly independent real solutions ν, and it follows that

$$b_0 = 2p - r. \tag{14}$$

We see that a_0 and b_0 are finite and satisfy the assertion (3). This completes the proof of the Riemann–Roch theorem.

The Riemann–Roch theorem is algebraic in nature since its formulation requires only the use of algebraic concepts; namely, algebraic functions, abelian differentials and ranks of systems of linear equations. Here we note that p can be defined as the number of linearly independent abelian differentials of the first kind. Accordingly, the theorem can be proved in a purely algebraic context and we need not assume that the field of constants is the field of complex numbers. Our present point of view precludes further study of this issue.

The familiar theorem on the partial fractions decomposition of a rational function is a special case of Theorem 2. Namely, let $\mathfrak{R}$ denote the schlicht number sphere. Consider all the rational functions of z with possible poles at $z_1 = \infty$ and at $n - 1$ distinct finite points $z_2, \ldots, z_n$ with maximal orders $l_1, \ldots, l_n$. The functions z^l $(l = 0, 1, \ldots, l_1)$ and $(z - z_k)^{-l}$ $(l = 1, \ldots, l_k;$ $k = 2, \ldots, n)$ are just such linearly independent functions. Since their number is $(l_1 + \cdots + l_k) + 1$ and $p = 0$, Theorem 2 implies that they form a basis.

Let $\mathfrak{R}$ again be arbitrary. For every integral divisor $\mathfrak{d}$ we have, invariably, $\mathfrak{d}^{-1} \mid 1$, and, therefore, in the statement of Theorem 2, $a \geqslant 1$. On the other hand, by Theorem 2 we also have $a = m - p + 1 + b \geqslant m - p + 1$, so that for $m > p$ surely $a > 1$. This means that for $m > p$ there are always nonconstant functions in K which are infinite at most at m arbitrarily assigned points.

Before proving the Riemann–Roch theorem for fractional divisors we present a few deeper applications of Theorem 2.

Theorem 3: The number of zeros of a differential of the first kind which does not vanish identically is $2p - 2$; that is, it is the same for all such differentials.

Proof: We suppose $p > 0$, for otherwise the assertion is meaningless. If du and dv are two nontrivial differentials of the first kind, then the corresponding divisors $(du) = \mathfrak{u}$ and $(dv) = \mathfrak{v}$ are integral and the quotient $du/dv = g$ is a function in K with, possibly reducible, associated divisor $\mathfrak{u}/\mathfrak{v} = (g)$. By Theorem 1, $\mathfrak{u}$ and $\mathfrak{v}$ have the same degree. This shows that the number m of zeros is the same for all differentials du of the first kind.

Furthermore, if dw is an abelian differential for which $\mathfrak{u} \mid dw$, then, in particular, $dw = 0$ or (dw) is integral, and so, in the latter case, $(dw) = \mathfrak{u} = (du)$, which implies that $dw = \lambda\, du$ for some constant λ. We see that the differentials dw in question form a vector space of dimension $b = 1$. Applying Theorem 2 with $\mathfrak{d} = \mathfrak{u}$ and considering the totality of functions f in K which satisfy the condition $\mathfrak{u}^{-1} \mid \mathfrak{f}$, we obtain for it the dimension

$$a = m - p + 2. \tag{15}$$

We now show that the functions under consideration are precisely the functions

$$f = \frac{dv}{du}, \tag{16}$$

with dv an arbitrary differential of the first kind and du with $(du) = \mathfrak{u}$ fixed. In fact, if f is given by (16), then $f = 0$ or $\mathfrak{u}(f)$ is integral, that is, $\mathfrak{u}^{-1} \mid f$. Conversely, if $\mathfrak{u}^{-1} \mid f$, then the differential $f\,du = dv$ is free of poles, and, therefore, of the first kind. Since p is the dimension for the differentials of the first kind, it follows that $a = p$ and, by (15), this proves our assertion.

The following theorem is a generalization of Theorem 3.

Theorem 4: Let α be the number of zeros and β the number of poles of a nontrivial differential. Then

$$\alpha - \beta = 2p - 2.$$

Proof: If du and dv are two not identially vanishing differentials, then their quotient du/dv is again a function in K. Then (g) has degree 0, which means that (du) and (dv) are of the same degree. This degree, however, is the difference $\alpha - \beta$ between the number of zeros and poles of the differential du. It follows that this difference has the same value for all differentials which do not vanish identically. If $p > 0$, then we can pick for du, in particular, a differential of the first kind and obtain the required assertion from Theorem 3. In the case $p = 0$, we must proceed differently. Since, by a remark subsequent to Theorem 2, for $m > p$, K must contain nonconstant functions with at most m poles, it follows that for $p = 0$ there is in K a function f of order 1. By Theorem 1, this function takes on $\mathfrak{R}$ every value exactly once, and, therefore, maps $\mathfrak{R}$ conformally onto the schlicht sphere. Since, for finite values of f, df is different from 0, and since, in view of $df = -f^2\, df^{-1}$, df has a double pole at the pole of f, we have, in this case, $\alpha = 0$ and $\beta = 2$; that is, again, $\alpha - \beta = -2 = 2p - 2$.

Without using Theorems 3 and 2 we can argue in all cases in the following manner, provided that we have already established the fact that for all differentials not vanishing identically the difference $\alpha - \beta$ has the same value. We consider, in particular, the differential dx of the independent variable x on $\mathfrak{R}$. If the exact number of sheets of $\mathfrak{R}$ joined locally at a finite point is j, then dx vanishes there to order $j - 1$. Similarly, every point at infinity is a pole of order $j + 1$. If n is the number of sheets of $\mathfrak{R}$, then n is at the same time the sum of the values of j associated with $x = \infty$. Using the branch number v of $\mathfrak{R}$ defined in Section 4 of Chapter 2 (Volume I), we have $\alpha - \beta = v - 2n$. On the other hand, we proved there that $2n - v = 2 - 2p$. This yields another proof of the theorem.

If the degree m of a given integral divisor $\mathfrak{d}$ is greater than $2p - 2$, then by Theorem 3 there is no differential dw other than 0 for which $\mathfrak{d} \mid dw$, and so $b = 0$. Now Theorem 2 implies the simple relation

$$a = m - p + 1 \qquad (m > 2p - 2) \tag{17}$$

regardless of the choice of $\mathfrak{d}$.

We have already noted that for $m > p$ we always have $a > 1$, so that for every integral divisor $\mathfrak{d}$ of degree m there is a nonconstant function f in K for which $\mathfrak{d}^{-1} \mid f$. In particular, there is always a nonconstant function in K which has poles in at most $p + 1$ arbitrarily given points. Now let $m \leqslant p$ and $\mathfrak{d} = \mathfrak{q}_1 \cdots \mathfrak{q}_m$. If we are to have $a > 1$, then, by Theorem 2, the condition

$$m - p + b > 0 \tag{18}$$

must hold. If dw is a differential with no pole at a point $\mathfrak{q}$ of $\mathfrak{R}$, then by $dw/d\mathfrak{q}$ we mean the value of $dw/d\zeta$ at $\zeta = 0$, where ζ is the local parameter. Higher derivatives are treated similarly. After choosing a basis $dw_1, \ldots, dw_p$ for the differentials of the first kind, we form the matrix

$$W(\mathfrak{d}) = \frac{dw_1}{d\mathfrak{q}_k} \tag{19}$$

of m rows and p columns. If the point $\mathfrak{q}_k$ turns up with multiplicity m_k, then in (19) the m_k rows in question are to be replaced by the derivatives of dw_l $(l = 1, \ldots, p)$ with respect to ζ of order up to m_k. Since the vanishing of the differential of the first kind

$$dw = \lambda_1\, dw_1 + \cdots + \lambda_p\, dw_p$$

at the point $\mathfrak{q}_k$ to an order $\geqslant m_k$ is expressed by the m_k equations

$$\lambda_1 \frac{d^s w_1}{(d\mathfrak{q}_k)^s} + \cdots + \lambda_p \frac{d^s w_p}{(d\mathfrak{q}_k)^s} = 0 \qquad (s = 1, \ldots, m_k),$$

the rank r of $W(\mathfrak{d})$ and the number b are connected by the formula

$$b + r = p.$$

It follows that condition (18) goes over into

$$r < m,$$

so that all the subdeterminants of $W(\mathfrak{d})$ of order m must vanish. It remains to show that this is not an empty restriction, that is, one which is satisfied identically in $\mathfrak{q}_1, \ldots, \mathfrak{q}_m$. Let $\mathfrak{q}_1, \ldots, \mathfrak{q}_m$ vary independently, and let us suppose that the subdeterminant of $W(\mathfrak{d})$ consisting of the first $r + 1$ rows and columns of $W(\mathfrak{d})$ is identically 0 but that this is not the case for all minors of order r of elements of the first row; here $r < m$. Since dw_1 is not identically 0, we certainly have $r \geqslant 1$. Expansion of our determinant by the first row would yield a contradiction, namely, the linear dependence of $dw_1, \ldots, dw_{r+1}$. The more general case, when the points $\mathfrak{q}_1, \ldots, \mathfrak{q}_m$ appear with certain multiplicities and otherwise vary independently, can be treated by making use of a certain property of the Wronskian. We shall not consider this issue further since no use is made of it in the sequel.

We now derive a number of consequences of Theorem 2 for the case $p = 1$. Let w be the global uniformizing parameter. By Theorem 2 in Section 7 of the preceding chapter, every covering transformation of $\mathfrak{R}$ has the effect of a translation on w. It follows that dw is single-valued and regular throughout $\mathfrak{R}$ and is, therefore, a differential of the first kind which, by Theorem 5, Section 2, is determined up to a constant factor. By Theorem 3, dw has no zeros, and this agrees with the fact that the mapping of $\mathfrak{R}$ onto the complex plane effected by w is conformal. On the other hand, Theorem 3 shows that for $p > 1$ every differential dw of the first kind has zeros, so that the integral w no longer effects an everywhere conformal mapping of the canonically dissected Riemann surface onto a schlicht region, which is why the integrals of the first kind cannot be used for purposes of uniformization. Let us again take $p = 1$ and let $\mathfrak{d} = \mathfrak{q}$ be any prime divisor. Since in that case $m = 1 > 2p - 2 = 0$, we have, by (17), $a = m - p + 1 = 1$, so that there is no function in K of order 1. This happens to be true for $p > 1$, as can be seen from Theorem 1 in Section 7 of Chapter 3. In order to obtain for $p = 1$ a simplest possible nonconstant function in K, we choose $\mathfrak{d} = \mathfrak{q}^2$, that is, $m = 2$, which implies $a = 2$. This means that there is a function x in K with $\mathfrak{q}^{-2} \mid x$ and $\mathfrak{q}^{-1} \nmid x$, and that the most general function f in K with $\mathfrak{q}^{-2} \mid f$ is $f = \lambda x + \mu$ with arbitrary constants λ, μ. Since dx is a differential, the quotient

$$y = \frac{dx}{dw} \tag{20}$$

is a function in K. Also, we can so normalize w that w vanishes at the point $\mathfrak{q}$. Then the following expansions hold in the vicinity of $\mathfrak{q}$:

$$x = cw^{-2} + \cdots, \qquad y = -2cw^{-3} + \cdots \qquad (c \neq 0).$$

Since dw is nowhere 0, $\mathfrak{q}$ is the only pole of y, and its order is three. By Theorem 1, y has exactly three zeros $\mathfrak{p}_1, \mathfrak{p}_2, \mathfrak{p}_3$, so that

$$(21) \qquad (y) = \frac{\mathfrak{p}_1\mathfrak{p}_2\mathfrak{p}_3}{\mathfrak{q}^3}.$$

Let e_k $(k = 1, 2, 3)$ be the value of x at the point $\mathfrak{p}_k$. If two of the points $\mathfrak{p}_k$ coincided, say, $\mathfrak{p}_1 = \mathfrak{p}_2$, then, by (20) and (21), the function $x - e_1$ would have at $\mathfrak{p}_1$ a zero of at least third order, while its order is two. This shows that $\mathfrak{p}_1, \mathfrak{p}_2, \mathfrak{p}_3$ are distinct and the difference

$$f_k = x - e_k \qquad (k = 1, 2, 3)$$

vanishes at $\mathfrak{p}_k$ to exactly the second order. It follows that

$$(22) \qquad (f_k) = \frac{\mathfrak{p}_k^2}{\mathfrak{q}^2} \qquad (k = 1, 2, 3),$$

so that the three numbers e_1, e_2, e_3 are also distinct. Equations (21) and (22) imply for the function

$$g = y^{-2}f_1f_2f_3$$

the relation $(g) = \mathfrak{e}$, so that g is a nonzero constant. Replacing x by $\lambda x + \mu$ with suitably chosen λ, μ, we can have

$$g = 1/4, \qquad e_1 + e_2 + e_3 = 0,$$

which would lead to the equation

$$y^2 = 4x^3 - g_2x - g_3$$

with constants g_2, g_3, and $g_2^3 - 27g_3^2 \neq 0$. Since this is an irreducible quadratic equation for y and, on the other hand, x has order two, x and y generate K. This proves anew a result stated in Theorem 1 of Section 8 of the preceding chapter.

Now we come to the general Riemann–Roch theorem for fractional divisors.

Theorem 5: *Theorem 2 holds for fractional divisors* $\mathfrak{d}$.

Proof: In spite of the fact that the proof is, in principle, a repetition of the proof of Theorem 2, the importance of the result compels us to present once more the three steps of the proof. We again put $2a = a_0$, $2b = b_0$.

I. Let

$$\mathfrak{d} = \frac{\mathfrak{d}_1}{\mathfrak{d}_2}, \qquad \mathfrak{d}_1 = \mathfrak{p}_1^{s_1} \cdots \mathfrak{p}_g^{s_g}, \qquad \mathfrak{d}_2 = \mathfrak{q}_1^{t_1} \cdots \mathfrak{q}_h^{t_h}$$

be the decomposition of $\mathfrak{d}$ into powers of different prime divisors. With the notations

$$s_1 + \cdots + s_g = s, \qquad t_1 + \cdots + t_h = t$$

the degree has the value

$$m = s - t. \tag{23}$$

The case $t = 0$ is settled by Theorem 2, so we may suppose $t > 0$. For the determination of the abelian integrals, we shift the initial point $\mathfrak{x}_0$ to $\mathfrak{q}_1$ so that, in particular, $E_l(\mathfrak{p}_k)$ and $F_l(\mathfrak{p}_k)$ vanish for $\mathfrak{x} = \mathfrak{q}_1$. In analogy with (5) we put

$$\varphi = \varphi(\mathfrak{x}) = \sum_{k=1}^{g} \sum_{l=1}^{s_k} (\lambda_{kl} F_l(\mathfrak{p}_k) + i\mu_{kl} E_l(\mathfrak{p}_k)), \tag{24}$$

so that $\varphi(\mathfrak{q}_1) = 0$. All the functions f in K with the property $\mathfrak{d}^{-1} \mid f$ are given by those integrals $\varphi(\mathfrak{x})$ defined in (24) which satisfy the following additional conditions:

$$\begin{aligned} &[\varphi] = 0, \qquad \varphi(\mathfrak{q}_k) = 0 \qquad (k = 2, \ldots, h), \\ &\frac{d^n\varphi}{(d\mathfrak{q}_k)^n} = 0, \qquad (n = 1, \ldots, t_k - 1; \qquad k = 1, \ldots, h). \end{aligned} \tag{25}$$

In the case $p = 0$ the first condition is meaningless.

We now come to a consideration which did not arise at the corresponding point in the proof of Theorem 2. We consider the totality of differentials dw with the property $\mathfrak{d}_2^{-1} \mid dw$. These differentials can be constructed out of elementary differentials as follows. In addition to the $2p$ differentials $dw_1, \ldots, dw_{2p}$ of the first kind linearly independent over the reals, we take the differentials $dE(\mathfrak{q}_k, \mathfrak{q}_1)$ and $i\, dE(\mathfrak{q}_k, \mathfrak{q}_1)$ $(k = 2, \ldots, h)$ of the third kind and the differentials $dE_n(\mathfrak{q}_k)$ and $i\, dE_n(\mathfrak{q}_k)$ $(n = 1, \ldots, t_k - 1;\ k = 1, \ldots, h)$ of the second kind. Their number is

$$2q = 2p + 2(h - 1) + 2(t - h) = 2p + 2t - 2. \tag{26}$$

By Theorem 4 in Section 1, all the differentials dw under consideration can be uniquely represented as real linear combinations of these differentials.

II. We suppose the curves of the canonical dissection of $\mathfrak{R}$ to bypass all the points $\mathfrak{p}_1, \ldots, \mathfrak{p}_g$ and $\mathfrak{q}_1, \ldots, \mathfrak{q}_h$. For each of the integrals φ and differentials dw introduced above the expression $\varphi\, dw$ is single-valued on the canonically dissected surface $\mathfrak{R}^*$, and regular apart from poles. It follows that for $p > 0$ we obtain once more the formula (8), where the left-hand side is equal to the sum of the residues of $\varphi\, dw/d\zeta$ on $\mathfrak{R}^*$. In the present case, however, we must consider the points $\mathfrak{q}_1, \ldots, \mathfrak{q}_h$. If, for $k = 1, \ldots, h$, φ vanishes at $\mathfrak{q}_k$ to at least order t_k, then $\varphi\, dw$ is regular there; and, as before in forming the residues we need concern ourselves only with the points $\mathfrak{p}_1, \ldots, \mathfrak{p}_g$. Then the real part of the sum of the residues is again given by the expression $\rho = \rho(\varphi\, dw)$ defined in (9). If, in addition, $[\varphi] = 0$, then (8) implies the equality $\rho = 0$. With the assumption (25) satisfied, this equality holds in the case $p = 0$ as well, for $\varphi\, dw$ is then an abelian differential whose residue sum is, by Theorem 1 of Section 1, equal to 0.

Now we assume, conversely, that for a fixed φ of the form (24) the condition $\rho = 0$ holds for all dw with the property $\mathfrak{d}_2^{-1} \mid dw$. If, to begin with, we take for dw, in particular, the differentials $dw_1, \ldots, dw_{2p}$, then we obtain, just as in step II of the proof of Theorem 2, the formula $[\varphi] = 0$. It follows that φ is in K, and that the residue sum of the abelian differential $\varphi\, dw$ is 0 for all dw under consideration. Since $\rho = 0$, the real part of the residue sum taken over the points $\mathfrak{q}_1, \ldots, \mathfrak{q}_h$ only is also 0. Putting successively for dw the differentials $dE(\mathfrak{q}_k, \mathfrak{q}_1)$, $i\, dE(\mathfrak{q}_k, \mathfrak{q}_1)$ $(k = 2, \ldots, h)$, $dE_n(\mathfrak{q}_k)$, $i\, dE_n(\mathfrak{q}_k)$ $(n = 1, \ldots, t_k - 1;\ k = 1, \ldots, h)$ and bearing in mind the equality $\varphi(\mathfrak{q}_1) = 0$, we see that the function φ satisfies all the conditions in (25) and, consequently, $\varphi = f$ has the required property $\mathfrak{d}^{-1} \mid f$.

III. Now we are given a fixed differential dw with the property $\mathfrak{d}_2^{-1} \mid dw$. In addition, the condition $\rho(\varphi\, dw) = 0$ holds for all integrals φ of the form (24). If we successively put in place of φ the integrals $F_l(\mathfrak{p}_k)$, $iE_l(\mathfrak{p}_k)$ $(l = 1, \ldots, s_k;\ k = 1, \ldots, g)$, this condition implies that, for $k = 1, \ldots, g$, dw vanishes at the point $\mathfrak{p}_k$ at least to order s_k. Conversely, this vanishing implies the equality $\rho = 0$ for all φ under consideration; also, it is equivalent to the assertion $\mathfrak{d} \mid dw$.

Now we form the matrix R with entries $\rho(\varphi\, dw)$, where for φ we take the $2s$ integrals $F_l(\mathfrak{p}_k)$, $iE_l(\mathfrak{p}_k)$ $(l = 1, \ldots, s_k;\ k = 1, \ldots, g)$ and for dw the $2q$ differentials dw_k $(k = 1, \ldots, 2p)$, $dE(\mathfrak{q}_k, \mathfrak{q}_1)$ and $i\, dE(\mathfrak{q}_k, \mathfrak{q}_1)$ $(k = 2, \ldots, h)$, $dE_n(\mathfrak{q}_k)$ and $i\, dE_n(\mathfrak{q}_k)$ $(n = 1, \ldots, t_k - 1;\ k = 1, \ldots, h)$. If R has rank r, then we have

$$2a = 2s - r, \qquad 2b = 2q - r.$$

Now (23) and (26) yield the assertion

$$a - b = s - q = (m + t) - (p + t - 1) = m - p + 1.$$

As an application of Theorem 5 we give a direct proof of Theorem 4. Let du be a differential which does not vanish identically, and let $(du) = \mathfrak{d}$ be of degree m; that is, m is the difference of the zeros α and the poles β of du. If, in addition, $\mathfrak{d} \mid dw$, then the quotient dw/du is a function in K regular throughout $\mathfrak{R}$ and, as such, is constant. Hence $b = 1$. By Theorem 5, the number of linearly independent f in K with $\mathfrak{d}^{-1} \mid f$ is

$$a = m - p + 1 + b = m - p + 2.$$

For every such f the differential $f\,du$ is of the first kind and, conversely, for every differential dv of the first kind the quotient $dv/du = f$ is an admissible function. It follows that $a = p$ and $m = 2p - 2$.

7. *The theorem of Abel*

In this section we discuss the question mentioned earlier of the conditions under which a divisor $\mathfrak{d}$ is a principal divisor, that is, a divisor for which there is a function f in K with $(f) = \mathfrak{d}$. The solution of this question will lead us subsequently to the addition theorem for integrals of the first kind.

By Theorem 1 of the preceding section, a not identically vanishing function f in K has as many zeros as poles, so that every principal divisor has degree 0. It is clear that the principal divisors form a subgroup of the group of all divisors. By forming the factor group we introduce a notion of equivalence and we write $\mathfrak{a} \sim \mathfrak{b}$ if the quotient

$$\mathfrak{d} = \frac{\mathfrak{a}}{\mathfrak{b}} \tag{1}$$

of the two divisors $\mathfrak{a}$ and $\mathfrak{b}$ is a principal divisor. The elements of the factor group are the classes of divisors, called divisor classes, corresponding to this notion of equivalence. Each of these classes is assigned a degree. The classes of degree 0 form a multiplicative group whose significance will be apparent in the sequel.

Let $\mathfrak{d}$ be any divisor (which we suppose in the form (1)) of a not necessarily reduced quotient of integral divisors $\mathfrak{a}$ and $\mathfrak{b}$. For $\mathfrak{d}$ to be a principal divisor, that is, for $\mathfrak{d} \sim \mathfrak{e}$, the degree m of $\mathfrak{a}$ must be equal to the degree of $\mathfrak{b}$; we can, therefore, write $\mathfrak{a}$ and $\mathfrak{b}$ as two products

$$\mathfrak{a} = \mathfrak{p}_1 \cdots \mathfrak{p}_m, \qquad \mathfrak{b} = \mathfrak{q}_1 \cdots \mathfrak{q}_m \tag{2}$$

of m not necessarily distinct prime divisors each. We start with a definite canonical dissection of the Riemann surface $\mathfrak{R}$ whose curves avoid the $2m$ points $\mathfrak{p}_1, \ldots, \mathfrak{q}_m$. The answer to the question posed above is given by the following result known as *Abel's theorem.*

Theorem 1: The relation

$$\mathfrak{p}_1 \cdots \mathfrak{p}_m \sim \mathfrak{q}_1 \cdots \mathfrak{q}_m \tag{3}$$

holds if and only if for every differential dw of the first kind, integration on the canonically dissected Riemann surface yields

$$\sum_{k=1}^{m} \int_{\mathfrak{q}_k}^{\mathfrak{p}_k} dw = w(L), \tag{4}$$

where L is a suitable closed curve on $\mathfrak{R}$ independent of the choice of w.

Proof: Let (3) hold and let $\mathfrak{d}$ be defined by (1) and (2), that is, let $\mathfrak{d}$ be a principal divisor. With f a function in K defined by $(f) = \mathfrak{d}$, and w an arbitrary integral of the first kind, $(w/2\pi i)\, d \log f$ is single-valued on the canonically dissected surface $\mathfrak{R}^*$. We integrate $(w/2\pi i)\, d \log f$ over the boundary C of $\mathfrak{R}^*$. The abelian differential $d \log f$ has as poles precisely those points $\mathfrak{r}$ which do not vanish under reduction of the right-hand side of (1); specifically, $\mathfrak{r}$ is a simple pole with residue l if the power $\mathfrak{r}^l$ remains on the right-hand side of (1). The corresponding residue of $w\, d \log f$ is $lw(\mathfrak{r})$, so that, by summing over all the poles $\mathfrak{r}$ we obtain as residue sum the value

$$\sum_{k=1}^{m} w(\mathfrak{p}_k) - \sum_{k=1}^{m} w(\mathfrak{q}_k) = \sum_{k=1}^{m} \int_{\mathfrak{q}_k}^{\mathfrak{p}_k} dw,$$

where we integrate on $\mathfrak{R}^*$. By a frequently used argument, the integral under consideration can also be obtained in the manner of the integral in (3), Section 3. Hence

$$\sum_{k=1}^{m} \int_{\mathfrak{q}_k}^{\mathfrak{p}_k} dw = [w]'J\left[\frac{1}{2\pi i} \log f\right]. \tag{5}$$

Since f is single-valued on $\mathfrak{R}$, the elements of the column $[1/2\pi i \log f]$ are integers $g_1, \ldots, g_p, h_1, \ldots, h_p$ which are independent of the choice of the integral w, but may depend on $\mathfrak{d}$ and on the given pairs of crosscuts A_k, B_k $(k = 1, \ldots, p)$. If we fix the homology class of a closed curve L by the condition

$$L \backsim \sum_{k=1}^{p} (h_k A_k - g_k B_k), \tag{6}$$

then we have

$$[w]'J\left[\frac{1}{2\pi i} \log f\right] = w(L), \tag{7}$$

and (4) follows from (6) and (7).

Conversely, let (4) hold for an appropriate curve L and for every differential dw of the first kind. Using the elementary integrals $E(\mathfrak{p}_k, \mathfrak{q}_k)$ $(k = 1, \ldots, m)$ of the third kind, we form the function

$$\psi = \sum_{k=1}^{m} E(\mathfrak{p}_k, \mathfrak{q}_k),$$

where in the case $\mathfrak{p}_k = \mathfrak{q}_k$, $E(\mathfrak{p}_k, \mathfrak{q}_k)$ is to be replaced by 0. If $\mathfrak{p}_k \neq \mathfrak{q}_k$, then the differential $dE(\mathfrak{p}_k, \mathfrak{q}_k)$ has simple poles at $\mathfrak{p}_k$ and $\mathfrak{q}_k$ with residues 1 and -1 and is otherwise regular on $\mathfrak{R}$. If we put

$$\psi = \log f.$$

then the function f is meromorphic on $\mathfrak{R}^*$ and has there precisely the zeros and poles prescribed by the divisor $\mathfrak{d}$. If it happens to be single-valued on $\mathfrak{R}$, then it is in K; so that $\mathfrak{d} \sim \mathfrak{e}$, and (3) follows. In the case $p = 0$, $\mathfrak{R}^* = \mathfrak{R}$, and the proof is complete. Let $p > 0$. By Theorem 2 of Section 2, the periods of ψ on $\mathfrak{R}$ are all pure imaginary and f will possess the required single-valuedness if and only if the real column $[(1/2\pi i)\psi]$ has integer entries exclusively. The reasoning in the preceding paragraph yields, in analogy to (5),

$$\sum_{k=1}^{m} \int_{\mathfrak{q}_k}^{\mathfrak{p}_k} dw = [w]'J\left[\frac{1}{2\pi i}\psi\right] \tag{8}$$

for every integral w of the first kind, the integration taking place on $\mathfrak{R}^*$. If γ is the column of the numbers $g_1, \ldots, g_p, h_2, \ldots, h_p$ determined uniquely by (6), then (4) and (8) imply the formula

$$[w]'J\left[\frac{1}{2\pi i}\psi\right] - \gamma = 0. \tag{9}$$

We apply it to the differentials $dw_1, \ldots, dw_{2p}$ constructed in the proof of Theorem 5 in Section 2, and note that the real part of the matrix $([w_1] \cdots [w_{2p}])$ is just the unit matrix E_{2p}. By forming the real part, we obtain from (9) the relation

$$\left[\frac{1}{2\pi i}\psi\right] = \gamma. \tag{10}$$

This completes the proof.

Formula (4) can be written somewhat differently if L is made to issue from $\mathfrak{p}_1$ and the path L^{-1} is adjoined to the path on $\mathfrak{R}^*$ from $\mathfrak{q}_1$ to $\mathfrak{p}_1$. Then (4) goes over into

$$\sum_{k=1}^{m} \int_{\mathfrak{q}_k}^{\mathfrak{p}_k} dw = 0, \tag{11}$$

where the m paths of integration need no longer be on $\mathfrak{R}^*$, but, independently of the choice of w, connect the given limits on $\mathfrak{R}$ in an appropriate manner. Using the definitions in (2), (11) can be written in abbreviated symbolic form as

$$\int_{\mathfrak{b}}^{\mathfrak{a}} dw = 0, \tag{12}$$

and this is the necessary and sufficient condition for the validity of the relation $\mathfrak{a} \sim \mathfrak{b}$.

In distinction to the Riemann–Roch theorem, Abel's theorem is transcendental in nature for it contains assertions about the values of abelian integrals. Abel himself proved only half of the theorem, namely, the necessity of condition (4) or (12) and this, too, he put in a different form and in a more general context. The full theorem was first formulated many decades later by Clebsch. For the special case $p = 1$, Abel's theorem appeared already in Section 14 of Chapter 1 (Volume I) (but see also Theorem 1 in Section 8 of Chapter 3). There is a close connection between Abel's theorem and the addition theorem for abelian integrals, a topic which we explore next. Before doing so, however, we must embark on some preliminary investigations. From now on we again suppose $p > 0$.

Let $\mathfrak{x}_1, \ldots, \mathfrak{x}_m$ be variable points on $\mathfrak{R}$, and let $\mathfrak{x}_1 \cdots \mathfrak{x}_m = \mathfrak{x}$. Let us hold the class of the divisor $\mathfrak{x}$ fixed and let the m points converge to limits $\mathfrak{a}_1, \ldots, \mathfrak{a}_m$. We wish to show that the limit divisor $\mathfrak{a} = \mathfrak{a}_1 \cdots \mathfrak{a}_m$ also belongs to that class. If $\mathfrak{x}_k$ varies over the sequence of points $\mathfrak{x}_{kn}$ ($n = 1, 2, \ldots$), then by Abel's theorem we have for every differential dw of the first kind

$$\sum_{k=1}^{m} \int_{\mathfrak{x}_{k1}}^{\mathfrak{x}_{kn}} dw = w(L_n) \qquad (n = 1, 2, \ldots), \tag{13}$$

where the integrations on the left are carried out on the canonically dissected surface $\mathfrak{R}^*$, and the curve L_n, closed on $\mathfrak{R}$, is independent of the choice of w. Now let

$$L_n \backsim \sum_{l=1}^{p} (h_{ln}A_l - g_{ln}B_l) \qquad (n = 1, 2, \ldots)$$

with integer coefficients h_{ln}, g_{ln}. Using (13) in the manner of (4) in the second part of the proof of Theorem 1, we see from (8) and (10) that the numbers h_{ln} and g_{ln} depend continuously on $\mathfrak{x}_{1n}, \ldots, \mathfrak{x}_{mn}$. This and the assumed convergence of the $\mathfrak{x}_{kn}$ show that for all sufficiently large n the numbers in question must have fixed values h_l and g_l. With

$$L \backsim \sum_{l=1}^{p} (h_l A_l - g_l B_l),$$

it follows that

$$\sum_{k=1}^{m} \int_{\mathfrak{x}_{k1}}^{\mathfrak{a}_k} dw = w(L).$$

From this the required assertion follows by Abel's theorem. We can, therefore, say that in every class the integral divisors form a closed set.

From now on let $m - 2p = n > 0$. Let $\mathfrak{q}_1, \ldots, \mathfrak{q}_m$ be m given distinct points on $\mathfrak{R}$, and let $\mathfrak{q}$ be the divisor $\mathfrak{q}_1 \cdots \mathfrak{q}_m$. Using p differentials of the first kind $dw_1, \ldots, dw_p$ which are linearly independent over the complex numbers, we form, as in (19) of the preceding section, the matrix

$$W(\mathfrak{q}) = \frac{dw_l}{d\mathfrak{q}_k}$$

with m rows and p columns. There we showed that none of the subdeterminants of order p of $W(\mathfrak{q})$ vanished identically in the variables $\mathfrak{q}_1, \ldots, \mathfrak{q}_m$. Now we choose the m points so that all those subdeterminants are different from 0. Next we apply the Riemann–Roch theorem with $\mathfrak{d} = \mathfrak{q}$. Since the degree m of $\mathfrak{q}$ is greater than $2p - 2$, (17) in Section 6 implies the relation

$$a = m - p + 1 = p + n + 1.$$

We choose a basis $f_0, f_1, \ldots, f_{p+n}$ for the functions f in K defined by the condition $\mathfrak{d}^{-1} \mid f$, and for each such f we then have

$$f = \lambda_0 f_0 + \lambda_1 f_1 + \cdots + \lambda_{p+n} f_{p+n} \tag{14}$$

with uniquely determined complex coefficients $\lambda_0, \lambda_1, \ldots, \lambda_{p+n}$. Now let $\mathfrak{x}_1, \ldots, \mathfrak{x}_{p+n}$ be $p + n$ arbitrary points on $\mathfrak{R}$ supposed distinct and different from $\mathfrak{q}_1, \ldots, \mathfrak{q}_m$. For all of them to be zeros of the function f, it is necessary and sufficient that the $p + n$ homogeneous linear equations

$$\lambda_0 f_0(\mathfrak{x}_k) + \cdots + \lambda_{p+n} f_{p+n}(\mathfrak{x}_k) = 0 \qquad (k = 1, \ldots, p + n) \tag{15}$$

in the $p + n + 1$ unknowns $\lambda_0, \ldots, \lambda_{p+n}$ are satisfied. We also suppose the determinant $|f_l(\mathfrak{x}_k)|$, $k = 1, \ldots, p + n$, $l = 1, \ldots, p + n$, of order $p + n$ to be different from 0, an assumption which certainly holds for indeterminate $\mathfrak{x}_1, \ldots, \mathfrak{x}_{p+n}$, owing to the linear independence of the functions $f_1, \ldots, f_{p+n}$. The relation (15) yields the $p + n$ ratios $\lambda_1/\lambda_0, \ldots, \lambda_{p+n}/\lambda_0$ as rational functions of the coordinates x_k, y_k of all the points $\mathfrak{x}_k$ ($k = 1, \ldots, p + n$) which clearly are invariant under arbitrary permutations of the $\mathfrak{x}_k$. We shall refer to such functions as *symmetric rational functions of the divisor* $\mathfrak{x}_1 \cdots \mathfrak{x}_{p+n}$ *or of the points* $\mathfrak{x}_1, \ldots, \mathfrak{x}_{p+n}$. If in (14) we choose $\lambda_0 = 1$, then f is uniquely determined, and, with $\mathfrak{x}_1 \cdots \mathfrak{x}_{p+n} = \mathfrak{x}$, we have

$$\frac{\mathfrak{x}}{\mathfrak{q}} \,\Big|\, f, \qquad (f) = \frac{\mathfrak{x}\mathfrak{t}}{\mathfrak{q}}, \qquad \mathfrak{t} = \mathfrak{t}_1 \cdots \mathfrak{t}_p,$$

where the p points $\mathfrak{t}_1, \ldots, \mathfrak{t}_p$ are the remaining zeros of f or factors of $\mathfrak{q}$. Since prescribing $\mathfrak{x}$ determines (f) uniquely, $\mathfrak{t}$ is the only integral divisor in its class.

Retaining these assumptions, we let the various points $\mathfrak{x}_k$ ($k = 1, \ldots, p + n$) converge to $\mathfrak{q}_k$ and choose on $\mathfrak{R}$ arbitrary limit points $\mathfrak{t}_1^*, \ldots, \mathfrak{t}_p^*$ of the respective $\mathfrak{t}_1, \ldots, \mathfrak{t}_p$. Since the divisor $\mathfrak{x}\mathfrak{t}$ remains in the fixed class of $\mathfrak{q}$, we can apply the result in the last but one paragraph and see that the divisor $\mathfrak{t}_1^* \cdots \mathfrak{t}_p^* = \mathfrak{t}^*$ must lie in the class of $\mathfrak{q}_{p+n+1} \cdots \mathfrak{q}_m$. In view of the assumptions on $W(\mathfrak{q})$, the determinant $|W(\mathfrak{q}_{p+n+1} \cdots \mathfrak{q}_m)|$ is different from 0 and, using the application of the Riemann–Roch theorem discussed in connection with (19) in Section 6, we conclude that $\mathfrak{t}^* = \mathfrak{q}_{p+n+1} \cdots \mathfrak{q}_m$.

This shows that under the considered limiting process $\mathfrak{x}_k \to \mathfrak{q}_k$ $(k = 1, \ldots p + n)$ the properly ordered points $\mathfrak{t}_k$ $(k = 1, \ldots, p)$ tend to $\mathfrak{q}_{p+n+k}$. It follows, in particular, that if the $\mathfrak{x}_k$ $(k = 1, \ldots, p + n)$ lie sufficiently close to the p points $\mathfrak{t}_k$, then the latter are distinct and this is all the more true of indeterminate $\mathfrak{x}_1, \ldots, \mathfrak{x}_{p+n}$. It remains to see whether $\mathfrak{t}_1, \ldots, \mathfrak{t}_p$ are also different from $\mathfrak{q}_1, \ldots, \mathfrak{q}_m$. If, with $\hat{\mathfrak{q}}$ a fixed divisor of degree $m - r < m$, we could have the reduction

$$\frac{\mathfrak{t}}{\mathfrak{q}} = \frac{\hat{\mathfrak{t}}}{\hat{\mathfrak{q}}}, \qquad (f) = \frac{\mathfrak{x}\hat{\mathfrak{t}}}{\hat{\mathfrak{q}}},$$

then, in view of $m - r \geqslant m - p > p$ and the assumption concerning $W(\mathfrak{q})$, $W(\hat{\mathfrak{q}})$ would also have rank p. Application of the Riemann–Roch theorem would therefore yield for the number of linearly independent f with $\hat{\mathfrak{q}} \mid f$ the value $m - r - p + 1 < m - p + 1$. This contradiction shows that, for indeterminate $\mathfrak{x}$, $\mathfrak{t}$ has no prime divisor in common with $\mathfrak{q}$. We show further that the matrix $(f_l(\mathfrak{x}_k))$, with $k = 1, \ldots, p + n$ and $l = 0, \ldots, p + n$, must retain its rank $p + n$ upon replacement of $\mathfrak{x}_{n+1}, \ldots, \mathfrak{x}_{p+n}$ by $\mathfrak{t}_1, \ldots, \mathfrak{t}_p$. If this were not the case then we could find a function g in K such that, with $\mathfrak{z} = \mathfrak{x}_{n+1} \cdots \mathfrak{x}_{p+n}$,

$$(g) = \frac{\mathfrak{x}_1 \cdots \mathfrak{x}_n \mathfrak{t}\mathfrak{z}^*}{\mathfrak{q}}, \qquad \mathfrak{z}^* \neq \mathfrak{z};$$

but then, for indeterminate $\mathfrak{x}_{n+1}, \ldots, \mathfrak{x}_{n+p}$, there is no further integral divisor in the class of $\mathfrak{z}$. It follows that, conversely, $\mathfrak{z}$ is uniquely determined by $\mathfrak{t}$. This shows that, for variable $\mathfrak{x}_1, \ldots, \mathfrak{x}_n$, $\mathfrak{t}$ may be viewed as an independent variable and $\mathfrak{z}$ as a dependent variable.

Let ν be the order of x, that is, the number of sheets of $\mathfrak{R}$. We form all power products $x^k y^l$ $(l = 0, \ldots, \nu - 1;\ k = 0, 1, \ldots)$ and choose finitely many distinct ones, and denote them, in some definite order, by $g_1, \ldots, g_s$. This sequence may contain the elements x and y. Now let

$$g = \mu_1 g_1 + \cdots + \mu_s g_s \tag{16}$$

with indeterminates $\mu_1, \ldots, \mu_s$ which, for a given f, can be assigned complex values so that f and g generate K. Since $\mathfrak{x}\mathfrak{t}$ and $\mathfrak{q}$ have no prime divisor in common, f has order m. Hence g satisfies an irreducible equation

$$h_0 g^m + h_1 g^{m-1} + \cdots + h_m = 0, \tag{17}$$

with $h_0, \ldots, h_m$ relatively prime polynomials in f whose coefficients depend on $\mu_1, \ldots, \mu_s$ and $\mathfrak{x}_1, \ldots, \mathfrak{x}_{p+n}$. Let v be the number of coefficients. We apply (17) at $v - 1$ independently varying points of $\mathfrak{R}$. By forming determinants, we obtain, corresponding to the determination above of $\lambda_1, \ldots, \lambda_{p+n}$, the ratios of the coefficients as rational functions of $\mu_1, \ldots, \mu_s$ and

x_k, y_k $(k = 1, \ldots, p + n)$ which, in addition, are symmetric in $\mathfrak{x}_1, \ldots, \mathfrak{x}_{p+n}$. We suppose the last $v - 1$ points to have been assigned complex coordinates x, y so that the coefficients in (17) are not all 0.

Let $x = \xi_k$, $y = \eta_k$ $(k = 1, \ldots, p)$ be the coordinates of the points $\mathfrak{t}_1, \ldots, \mathfrak{t}_p$. For $f = 0$, (17) has the $p + n$ roots $g(x_k, y_k)$ $(k = 1, \ldots, p + n)$ and the p additional roots $g(\xi_k, \eta_k)$ $(k = 1, \ldots, p)$. Hence

$$\prod_{k=1}^{p} g(\xi_k, \eta_k) = \frac{(-1)^m h_m}{h_0 \prod_{k=1}^{p+n} g(x_k, y_k)}, \tag{18}$$

where we must put $f = 0$ in h_m and h_0. To begin with, the right-hand side is a rational function of $\mu_1, \ldots, \mu_s$. Since, by (16), the left-hand side is a polynomial in $\mu_1, \ldots, \mu_s$, it must be possible to reduce the right-hand side. Let $s_1, s_2, \ldots, s_p$ be any p not necessarily distinct numbers of the sequence from 1 to s. If we multiply the p factors on the left-hand side of (18), then the coefficient of $\mu_{s_1} \mu_{s_2} \cdots \mu_{s_p}$ is the sum of all the polynomials obtained from $g_{s_1}(\xi_1, \eta_1) g_{s_2}(\xi_2, \eta_2) \cdots g_{s_p}(\xi_p, \eta_p)$ by application of all the permutations of $s_1, s_2, \ldots, s_p$. However, for a suitable choice of $g_1, \ldots, g_s$, we can express every given symmetric rational integral function of the pairs ξ_k, η_k $(k = 1, \ldots, p)$ as a linear combination of these polynomials, and every symmetric rational function of $\mathfrak{t}_1, \ldots, \mathfrak{t}_p$ as a quotient of such linear combinations. We conclude, by (18), that every symmetric rational function of $\mathfrak{t}_1, \ldots, \mathfrak{t}_p$ is at once a rational symmetric function of $\mathfrak{x}_1, \ldots, \mathfrak{x}_{p+n}$.

After these preliminaries we can formulate the addition theorem for abelian integrals as follows.

Theorem 2: Let $\mathfrak{x} = \mathfrak{x}_1 \cdots \mathfrak{x}_n$, $\mathfrak{y} = \mathfrak{y}_1 \cdots \mathfrak{y}_n$ be two integral divisors of equal degree n, and let $\mathfrak{z} = \mathfrak{z}_1 \cdots \mathfrak{z}_p$ be an integral divisor of degree p. Then there exists an integral divisor $\mathfrak{v} = \mathfrak{v}_1 \cdots \mathfrak{v}_p$ of degree p such that, for appropriate choice of paths of integration for all the differentials dw of the first kind, we have the equality

$$\sum_{k=1}^{n} \int_{\mathfrak{y}_k}^{\mathfrak{x}_k} dw = \sum_{k=1}^{p} \int_{\mathfrak{z}_k}^{\mathfrak{v}_k} dw.$$

Furthermore, for undetermined $\mathfrak{x}$, $\mathfrak{y}$, $\mathfrak{z}$, every symmetric rational function of $\mathfrak{v}$ is a symmetric rational function of $\mathfrak{x}\mathfrak{z}$ and a symmetric rational function of $\mathfrak{y}$.

Proof: We make use of the general Riemann–Roch theorem with $\mathfrak{d} = \mathfrak{x}\mathfrak{z}/\mathfrak{y}$. Then $m = p$ and $a = m - p + 1 + b = b + 1 \geqslant 1$. Hence there exists a nontrivial function f in K with $\mathfrak{d}^{-1} \mid f$ and we have

$$f = \mathfrak{d}^{-1}\mathfrak{v} = \frac{\mathfrak{y}\mathfrak{v}}{\mathfrak{x}\mathfrak{z}}$$

with an integral divisor $\mathfrak{v}$ of degree p whose class is uniquely determined by the classes of $\mathfrak{x}$, $\mathfrak{y}$, $\mathfrak{z}$. By Abel's theorem and the remark connected with (11) and (12), the relation

$$\mathfrak{x}\mathfrak{z} \sim \mathfrak{y}\mathfrak{v}$$

implies that, after a suitable choice of path of integration for every abelian differential dw of the first kind,

$$\int_{\mathfrak{y}}^{\mathfrak{y}\mathfrak{v}} dw = 0,$$

and, consequently,

$$\int_{\mathfrak{y}}^{\mathfrak{x}} dw = \int_{\mathfrak{z}}^{\mathfrak{v}} dw. \tag{20}$$

This proves the first part of the theorem.

The fixed divisor $\mathfrak{q}$ is now to have its earlier meaning while $\mathfrak{x}\mathfrak{z}$ takes the place of the divisor which we denoted earlier by $\mathfrak{x}$. We form the integral divisor $\mathfrak{t} = \mathfrak{t}_1 \cdots \mathfrak{t}_p$ of degree p which is uniquely determined by the formula

$$\mathfrak{x}\mathfrak{z}\mathfrak{t} \sim \mathfrak{q}$$

with variable $\mathfrak{x}$, $\mathfrak{z}$; and then we form the uniquely determined integral divisor $\mathfrak{v} = \mathfrak{v}_1 \cdots \mathfrak{v}_p$ satisfying the condition

$$\mathfrak{y}\mathfrak{t}\mathfrak{v} \sim \mathfrak{q}$$

with variable $\mathfrak{y}$. In this way (19) is also satisfied, so that $\mathfrak{v}$ has the above meaning. By the earlier argument, every symmetric rational function of $\mathfrak{v}$ is symmetric rational in $\mathfrak{y}\mathfrak{t}$, and, similarly, every symmetric rational function of $\mathfrak{t}$ is symmetric rational in $\mathfrak{x}\mathfrak{z}$. This implies the rest of the theorem.

If we choose definite points on $\mathfrak{R}$ for the prime divisors in $\mathfrak{x}$, $\mathfrak{y}$, $\mathfrak{z}$, then it may happen that a given symmetric rational function of $\mathfrak{v}$ is of the form 0/0. This is why we considered variable $\mathfrak{x}$, $\mathfrak{y}$, $\mathfrak{z}$ and carried out the above preliminary investigations. These preliminary investigations also yield a constructive procedure for obtaining the rational functions under consideration.

An important special case of Theorem 2 is obtained for $n = p$ by addition of the identity

$$\int_{\mathfrak{y}}^{\mathfrak{z}} dw = \int_{\mathfrak{y}}^{\mathfrak{z}} dw \tag{21}$$

to (20). This leads to the formula

$$\int_{\mathfrak{y}}^{\mathfrak{x}} dw + \int_{\mathfrak{y}}^{\mathfrak{z}} dw = \int_{\mathfrak{y}}^{\mathfrak{v}} dw. \tag{22}$$

Here it is again the second assertion of Theorem 2 which is the main issue. In the case $p = 1$ this is just the addition theorem for elliptic functions of the first kind, and (22) is thus the generalization for any $p > 0$. If we interchange in (22) $\mathfrak{y}$ and $\mathfrak{z}$, replace $\mathfrak{v}$ by $\mathfrak{u}$, and add (21), then we get

$$(23) \qquad \int_{\mathfrak{y}}^{\mathfrak{x}} dw - \int_{\mathfrak{y}}^{\mathfrak{z}} dw = \int_{\mathfrak{y}}^{\mathfrak{u}} dw, \qquad \mathfrak{x}\mathfrak{y} \sim \mathfrak{z}\mathfrak{u}.$$

Thus every symmetric rational function of $\mathfrak{u}$ is symmetric rational in $\mathfrak{x}\mathfrak{y}$ and in $\mathfrak{z}$.

The relations (22) and (23) suggest that we associate with the divisor classes of degree p a commutative group written additively to distinguish it from the multiplicative group of the divisor classes, as follows. To begin with, we see, just as in the beginning of the proof of Theorem 2, that in every divisor class of degree p there is at least one integer divisor. Let $\{\mathfrak{x}\}$ be the divisor class of $\mathfrak{x}$ and let $\{\mathfrak{y}\}$ be a fixed class. For integral divisors $\mathfrak{x}, \mathfrak{y}, \mathfrak{z}$ of degree p, we now put

$$\{\mathfrak{x}\} + \{\mathfrak{z}\} = \{\mathfrak{v}\}, \qquad \{\mathfrak{x}\} - \{\mathfrak{z}\} = \{\mathfrak{u}\},$$

where the classes of the integral divisors $\mathfrak{u}$ and $\mathfrak{v}$ are uniquely determined by

$$\mathfrak{x}\mathfrak{z} \sim \mathfrak{y}\mathfrak{v}, \qquad \mathfrak{x}\mathfrak{y} \sim \mathfrak{z}\mathfrak{u}.$$

The zero element of this abelian group is clearly the fixed class $\{\mathfrak{y}\}$. In virtue of formulas (22) and (23), the symbolic addition and subtraction of divisor classes of degree p is expressed in the actual addition and subtraction of the associated integral sums.

8. The Jacobi inversion problem

By way of resuming the investigations of Chapter 1 (Volume I), we now concern ourselves with the inverse functions of the integrals of the first kind. In the case $p = 1$, the integral of the first kind is determined to within a constant factor. We can write it in the Weierstrass normal form

$$(1) \qquad s = \int_{\infty}^{x} \frac{dx}{y},$$

where the path of integration should be taken on the Riemann surface of the algebraic function $y = y(x)$ defined by the equation

$$y^2 = 4x^3 - g_2 x - g_3 \qquad (g_2^3 - 27g_3^2 \neq 0).$$

The inverse function of the function s of x defined by (1) is the elliptic $\wp$-function, so that

$$x = \wp(s), \qquad y = \frac{d\wp(s)}{ds}.$$

The $\wp$-function has two fundamental periods ω_1 and ω_2 such that each of its periods admits of a unique representation $\omega = m\omega_1 + n\omega_2$ with integer coefficients m and n. The existence of the fundamental periods is bound up with the fact that every Riemann surface of genus 1 has a basis in the sense of homology consisting of exactly two closed curves.

For $p > 1$ we could try, at first, to investigate the inverse function of a nontrivial integral w of the first kind. If $\mathfrak{a}_1$ and $\mathfrak{x}_1$ are two points on the Riemann surface $\mathfrak{R}$ with coordinates a_1, b_1 and x_1, y_1 in the independent variable x and the dependent variable y, then we put

$$s = \int_{\mathfrak{a}_1}^{\mathfrak{x}_1} dw$$

for fixed $\mathfrak{a}_1$ and variable $\mathfrak{x}_1$, and consider, conversely, x_1 as a function of s. Jacobi already surmised that, in distinction to the case $p = 1$, this inverse function could not be a meromorphic function of the variable s. Since the homology group of $\mathfrak{R}$ has a basis with $2p$ generators, we might expect the inverse function $x_1(s)$ to have $2p$ fundamental periods which are linearly independent over the field of rational numbers. We shall see in Chapter 5 that a nonconstant meromorphic function of one variable can have at most two such independent periods. In this line of reasoning we run into a difficulty when trying to establish the linear independence of the $2p$ periods of $x_1(s)$. It is far simpler to invoke Theorem 3 in Section 6, which states that the differential dw has, in all, $2p - 2$ zeros on $\mathfrak{R}$; for this implies that the inverse function $x_1(s)$ must have a branch point at each of the corresponding points of the s-plane.

Since there are p integrals $w_1, \ldots, w_p$ which are linearly independent over the complex numbers, one could attempt to generalize the construction of an inverse function carried out for $p = 1$ by inverting p suitable functions of p independent variables each. This idea occurred first to Jacobi, and was subsequently carried out for $p = 2$ by Göpel and Rosenhain. The statement is suggested by the form of Abel's theorem and the addition theorem. Given a fixed integral divisor $\mathfrak{a} = \mathfrak{a}_1 \cdots \mathfrak{a}_p$ of degree p, we require an unknown integral divisor $\mathfrak{x} = \mathfrak{x}_1 \cdots \mathfrak{x}_p$ of degree p such that, with p independent complex variables $s_1, \ldots, s_p$, we have the equalities

$$\int_{\mathfrak{a}}^{\mathfrak{x}} dw_l = \sum_{k=1}^{p} \int_{\mathfrak{a}_k}^{\mathfrak{x}_k} dw_l = s_l \qquad (l = 1, \ldots, p).$$

In the case $p = 1$ we come back to the inversion of equation (1). Closer investigation will show that the new formulation of the problem is reasonable and leads to interesting functions of p complex variables. To begin with, we shall prove the existence of a solution $\mathfrak{x}$ for arbitrarily assigned values of $s_1, \ldots, s_p$, and then we shall investigate more closely the dependence of the coordinates x_k, y_k $(k = 1, \ldots, p)$ of the p points $\mathfrak{x}_k$ on the variables $s_1, \ldots, s_p$. As suggested by the considerations in the previous paragraph, we shall investigate not the individual coordinates, but rather the symmetric rational functions of $\mathfrak{x}$ such as, say, $x_1 + \cdots + x_p$ and $x_1y_1^2 + \cdots + x_py_p^2$. It will turn out that these functions, viewed as functions of $s_1, \ldots, s_p$, are meromorphic functions with $2p$ independent periods. In the present section we clarify questions of existence and uniqueness of the divisor $\mathfrak{x}$. The remaining sections of this chapter are devoted to function-theoretical investigation of the inverse functions proper, that is, the symmetric rational functions $\mathfrak{x}$ viewed as functions of $s_1, \ldots, s_p$.

We shall refer to the problem of determining $\mathfrak{x}$ for given values of $s_1, \ldots, s_p$ as the *Jacobi inversion problem* for integrals of the first kind, or simply, as the *inversion problem*. It must be remembered that in Section 13 of Chapter 1 (Volume I), in connection with the investigation of the elliptic functions, the inversion problem consisted in finding an elliptic integral of the first kind with preassigned arbitrary fundamental periods. While this problem admits of a natural generalization to the case $p > 1$, it leads, in the general case, to unsolved difficulties; we shall not go into this matter here. From now on the term *inversion problem* will be used exclusively in the new sense.

Let $\mathfrak{q} = \mathfrak{q}_1 \cdots \mathfrak{q}_m$ be an integral divisor of degree m. We shall call it *general* if there are no nonconstant functions f in the field K with the property $\mathfrak{q}^{-1} \mid f$. Put differently, this means that $\mathfrak{q}$ is the only integral divisor in its class. By the Riemann–Roch theorem, the degree m of a general divisor is invariably $\leqslant p$. We shall call an integral divisor of degree $m \leqslant p$ which is not general a *special divisor*. It is clear from this definition that if $\mathfrak{q}$ is a special divisor, then so is every integral divisor which is equivalent to $\mathfrak{q}$.

In Section 6, as an application of Theorem 2, we found a necessary and sufficient condition for $\mathfrak{q}$ to be general. If the points $\mathfrak{q}_1, \ldots, \mathfrak{q}_m$ are all distinct, then we form the matrix $W(\mathfrak{q})$ of m rows and p columns whose elements $dw_l/d\mathfrak{q}_k$ are the derivatives relative to the respective local parameters at $\mathfrak{q}_k$ of the p integrals $w_1, \ldots, w_p$ of the first kind which are linearly independent over the complex numbers. If $\mathfrak{q}_1, \ldots, \mathfrak{q}_m$ are not all distinct, then let $\mathfrak{q}_1, \ldots, \mathfrak{q}_n$ be distinct and let their multiplicities be $g_1, \ldots, g_n$. Then the various columns of $W(\mathfrak{q})$ are formed out of the derivatives $d^g w_l/(d\mathfrak{q}_k)^g$ $(g = 1, \ldots, g_k;\ k = 1, \ldots, n)$. The divisor $\mathfrak{q}$ is general if and only if $W(\mathfrak{q})$ has the maximal rank m. This implies directly that if $\mathfrak{q}$ is general, then so is every divisor $\mathfrak{q}_1^{h_1} \cdots \mathfrak{q}_n^{h_n}$ $(0 \leqslant h_k \leqslant g_k;\ k = 1, \ldots, n)$. On the other hand, if $\mathfrak{q}$ is special then $W(\mathfrak{q})$ must have rank $r < m$. In that case, as we are about to show, the number r depends only on the class of $\mathfrak{q}$. As was shown in Section 6, $m - r + 1$ is the number of linearly independent

f in K with the property $\mathfrak{q}^{-1} \mid f$. If

$$\mathfrak{q} \sim \mathfrak{r}, \qquad \frac{\mathfrak{q}}{\mathfrak{r}} = (g),$$

then

$$\mathfrak{r} = \mathfrak{q}^{-1}(g) \mid fg,$$

so that, in fact, the number in question depends only on the class $\{\mathfrak{q}\}$. Furthermore, for any $m - r$ prescribed points $\mathfrak{y}_1, \ldots, \mathfrak{y}_{m-r}$, there are r additional points $\mathfrak{y}_{m-r+1}, \ldots, \mathfrak{y}_m$ such that

$$\text{(2)} \qquad \mathfrak{q} \sim \mathfrak{y}_1 \cdots \mathfrak{y}_{m-r}\mathfrak{y}_{m-r+1} \cdots \mathfrak{y}_m.$$

If we introduce a basis $f_0, f_1, \ldots, f_{m-r}$ for the f, assume that $\mathfrak{y}_1, \ldots, \mathfrak{y}_{m-r}$ are different from $\mathfrak{q}_1, \ldots, \mathfrak{q}_m$, and suppose the matrix with elements $f_l(\mathfrak{y}_k)$ $(k = 1, \ldots, m - r;\ l = 0, \ldots, m - r)$ to have rank $m - r$, then the divisor $\mathfrak{y}_{m-r+1} \cdots \mathfrak{y}_m$ in (2) is uniquely determined by $\mathfrak{y}_1 \cdots \mathfrak{y}_{m-r}$. This provides us with a survey of the integral divisors in the class $\{\mathfrak{q}\}$. We shall call r the *rank* of $\mathfrak{q}$, and of $\{\mathfrak{q}\}$. We note that for $p = 1$ the differential dw_1 has no zeros, and so every primitive divisor is general.

From now on we presuppose a canonical dissection of $\mathfrak{R}$, and with it a normal basis $w_1, \ldots, w_p$ of the integrals of the first kind with period matrix $Z = X + iY$. Then Z is symmetric and the imaginary part $Y > 0$. We put the p complex variables $s_1, \ldots, s_p$ in a column s and regard it as a point in the space of p complex coordinates. Similarly, w no longer denotes an arbitrary integral of the first kind, but rather the row with entries $w_1, \ldots, w_p$. The $2p$ rows of the full period matrix

$$Q = \begin{pmatrix} E \\ Z \end{pmatrix}$$

span in our complex p-dimensional space a parallelotope $\mathfrak{P}$ whose points are given by

$$\text{(3)} \qquad s = (\xi\eta)Q = \xi + \eta Z, \qquad \xi = (\xi_1 \cdots \xi_p), \qquad \eta = (\eta_1 \cdots \eta_p),$$

with $0 \leqslant \xi_k < 1$, $0 \leqslant \eta_k < 1$ $(k = 1, \ldots, p)$. Here we note that if $s = a + ib$ with real a, b, then, conversely, (3) implies

$$\eta = bY^{-1}, \qquad \xi = a - \eta X.$$

This establishes a one-to-one correspondence between the real space of the $2p$ coordinates $\xi_1, \ldots, \xi_p, \eta_1, \ldots, \eta_p$ and the complex space of the p coordinates $s_1, \ldots, s_p$. If, in particular, the $2p$ magnitudes $\xi_1, \ldots, \eta_p$ vary independently over the integers, then the corresponding points or periods $s = \omega$ form a lattice known as the *period lattice* Ω. If the difference $s - t$ of two points s and t is a period ω, then we call them *congruent* and write

$s \equiv t$. It is clear that every point s is congruent to just one point of $\mathfrak{P}$, and this makes $\mathfrak{P}$ into a fundamental region for the group of translations $s \to s + \omega$. In the case $p = 1$, $\mathfrak{P}$ reduces to the period parallelogram introduced in Chapter 1. We could have based the construction of $\mathfrak{P}$ and Ω on an arbitrary period matrix, but we chose instead a normalized period matrix as more appropriate for our purposes.

Concerning the solution of the Jacobi inversion problem, we have the following fundamental theorem.

Theorem 1: Let s be any point in the space of p complex variables. Then there exists an integral divisor $\mathfrak{x} = \mathfrak{x}_1 \cdots \mathfrak{x}_p$ of degree p such that

$$\int_{\mathfrak{a}}^{\mathfrak{x}} dw \equiv s \tag{4}$$

for every choice of the path of integration on $\mathfrak{R}$; thus the class of $\mathfrak{x}$ is uniquely determined.

Proof: Let $\mathfrak{b} = \mathfrak{b}_1 \cdots \mathfrak{b}_p$ be a general divisor of degree p with distinct points $\mathfrak{b}_1, \ldots, \mathfrak{b}_p$. To find such a divisor, we need only choose the distinct points $\mathfrak{b}_1, \ldots, \mathfrak{b}_p$ so that the determinant $|W(\mathfrak{b})| \neq 0$. If we put

$$t = s - \int_{\mathfrak{a}}^{\mathfrak{b}} dw,$$

then condition (4) is equivalent to

$$\int_{\mathfrak{b}}^{\mathfrak{x}} dw \equiv t.$$

It follows that in proving Theorem 1 we may suppose $\mathfrak{a}$ general and $\mathfrak{a}_1, \ldots, \mathfrak{a}_p$ distinct.

If, in addition to $\mathfrak{x}$, another integral divisor $\mathfrak{y}$ is a solution of (4), then by subtraction we obtain the condition

$$\int_{\mathfrak{y}}^{\mathfrak{x}} dw \equiv 0.$$

By Abel's theorem, this congruence implies that $\mathfrak{x} \sim \mathfrak{y}$ and conversely. It remains to prove the solvability of (4).

If $s = 0$, then $\mathfrak{x} = \mathfrak{a}$ is a solution and, in fact, $\mathfrak{a}$ being general, the only solution. Now we prove the existence of a unique solution $\mathfrak{x}$ when s is sufficiently close to 0, that is, when $|s_k| < \varepsilon$ $(k = 1, \ldots, p)$ for a sufficiently small positive ε. In fact, if $\zeta_1, \ldots, \zeta_p$ are the local parameters at $\mathfrak{a}_1, \ldots, \mathfrak{a}_p$, then the p functions

$$s_k = \sum_{l=1}^{p} \int_{\mathfrak{a}_l}^{\mathfrak{x}_l} dw_k = \sum_{l=1}^{p} \{w_k(\mathfrak{x}_l) - w_k(\mathfrak{a}_l)\} \qquad (k = 1, \ldots, p) \tag{5}$$

of $\zeta_1, \ldots, \zeta_p$ are single-valued and regular in a neighborhood of $\zeta_1 = 0, \ldots, \zeta_p = 0$; here the integrations are supposed rectilinear. Since the functional determinant

$$\left|\frac{ds_k}{d\zeta_l}\right| = \left|\frac{dw_k(\mathfrak{x}_l)}{d\zeta_l}\right| = |W(\mathfrak{x})|$$

is different from 0 at $\mathfrak{x} = \mathfrak{a}$, we conclude, by the existence theorem for inverse functions in the complex domain, that there is exactly one system of solutions $\zeta_k = \zeta_k(s_1, \ldots, s_p)$ of equations (5) which is regular in a sufficiently small neighborhood of $s_1 = 0, \ldots, s_p = 0$, and vanishes at $s = 0$. It follows that for a sufficiently small ε the determinant $|W(\mathfrak{x})| \neq 0$, that is, $\mathfrak{x}$ is general. This proves the unique solvability of (4) in a neighborhood of $s = 0$.

Now let s be arbitrary. Then we choose a natural number n so large that $|s/n| < \varepsilon$. In view of what has already been established, there is an integral divisor $\mathfrak{z}$ of degree p such that

$$\int_{\mathfrak{a}}^{\mathfrak{z}} dw \equiv \frac{s}{n}, \qquad \int_{\mathfrak{a}^n}^{\mathfrak{z}^n} dw = n \int_{\mathfrak{a}}^{\mathfrak{z}} dw \equiv s,$$

where, incidentally, $\mathfrak{z}$ is uniquely determined by the choice of n. By the addition theorem there is an integral divisor $\mathfrak{x}$ of degree p for which

$$\int_{\mathfrak{a}^n}^{\mathfrak{z}^n} dw = \int_{\mathfrak{a}}^{\mathfrak{x}} dw.$$

Clearly, requirement (4) is fulfilled. This completes the proof.

The divisor $\mathfrak{x}$ is a unique solution of the inversion problem if and only if it is general. On the other hand, if $\mathfrak{x}$ is special and of rank r, then we can assign $p - r$ of the p points $\mathfrak{x}_1, \ldots, \mathfrak{x}_p$ arbitrarily, say $\mathfrak{x}_k = \mathfrak{y}_k$ $(k = 1, \ldots, p - r)$. The remaining r points $\mathfrak{x}_{p-r+1}, \ldots, \mathfrak{x}_p$ are uniquely determined up to order if the matrix $(f_l(\mathfrak{y}_k))$ introduced earlier has rank $p - r$. It follows that the solution of the inversion problem depends on $p - r$ free parameters. In the sequel we shall call s *general* or *special of rank* r if the respective properties hold for the class $\{\mathfrak{x}\}$. Here we note that in this definition the class of the divisor $\mathfrak{a}$ must remain fixed. In the case $p = 1$, every divisor of degree p is prime and general, so that the inversion problem is invariably uniquely solvable. It follows that for $p = 1$ the canonically dissected Riemann surface is mapped by the integral $w(\mathfrak{x})$ of the first kind conformally onto a fundamental region relative to the period group Ω; this constitutes another proof of a result in Section 10 of Chapter 1 (Volume I). In that chapter the inversion problem was solved for $p = 1$ by the elliptic functions. In the following sections this problem will be solved for $p > 1$ by the introduction of the theta functions and the abelian functions. In particular, for $p = 1$, we shall obtain an important addition to the investigations of Chapter 1

(Volume I) in the form of a discussion of the relation between the elliptic functions and the theta functions.

By Theorem 1, there is a one-to-one correspondence between the points s in the period parallelogram $\mathfrak{P}$ and the classes $\{\mathfrak{x}\}$ of equivalent integral divisors of degree p. It is natural to regard the collection of these classes as a space and to introduce in it neighborhoods by first defining such neighborhoods in the complex p-dimensional space of the points s. The resulting topological space of the classes $\{\mathfrak{x}\}$ is called the *Jacobian variety*. In the case $p = 1$ we obtain in this way nothing new, since we are dealing with a Riemann surface of genus 1.

9. Theta functions

The theory of elliptic functions suggests that for $p > 1$ we look for meromorphic functions $f(s)$ of p complex variables $s_1, \ldots, s_p$ with Ω as period lattice, that is, for functions satisfying the equality $f(s + \omega) = f(s)$ identically in s for all ω in Ω. To obtain such functions explicitly, we first introduce the theta functions. By way of preparation we must investigate the convergence of infinite series to be introduced in the sequel. In the future we shall find it useful to view s not as a row but rather as a column with entries $s_1, \ldots, s_p$. We shall treat the periods ω similarly. Moreover, we shall use a column t of p real variables $t_1, \ldots, t_p$.

We consider a polynomial $g(t)$ of degree two in $t_1, \ldots, t_p$, that is,

$$g(t) = \varphi(t) + \psi(t) + c,$$

where $\varphi(t)$ is a quadratic form, $\psi(t)$ is a linear form, and c is a constant. The coefficients are complex numbers, and φ^*, ψ^*, c^*, denote the real parts of φ, ψ, c. Now let t vary in some given order over the lattice points of the real space of p dimensions so that, in effect, the coordinates $t_1, \ldots, t_p$ vary independently over the integers. We assume that, with

$$a(t) = e^{g(t)}, \tag{1}$$

the infinite series

$$\rho = \sum_t a(t) \tag{2}$$

converges under the given ordering of the t. Then $a(t)$ must tend to 0, that is, for a preassigned real T, the inequality

$$\varphi^*(t) + \psi^*(t) < T$$

must hold for almost all t of the sequence. Here we can replace t by $-t$ and obtain, accordingly,

$$\varphi^*(t) - \psi^*(t) < T, \tag{3}$$
$$\varphi^*(t) < T.$$

Given an integer $t \neq 0$, then, by replacing t by nt ($n = 1, 2, \ldots$), we obtain from (3) the inequality $\varphi^*(t) < Tn^{-2}$ for almost all n. Hence

$$\varphi^*(t) \leqslant 0. \tag{4}$$

In view of the homogeneity of φ^*, (4) holds for arbitrary rational t and, by continuity, for all real t. If there were a real $t = t_0 \neq 0$ with $\varphi^*(t_0) = 0$, then $\varphi^*(t)$ would have a maximum at t_0; for $t = t_0$, the polar $\varphi^*(t, u)$ of $\varphi^*(t)$ vanishes identically in u; and, for a scalar λ, we would have

$$\varphi^*(\lambda t_0 + u) = \lambda^2\varphi^*(t_0) + 2\lambda\varphi^*(t_0, u) + \varphi^*(u) = \varphi^*(u). \tag{5}$$

Now, for every positive integral value of λ we can find real values $u_1, \ldots, u_p$ of u such that $0 \leqslant u_k < 1$ and such that $\lambda t_0 + u = t$ is an integer. Since $\varphi^*(u)$ stays bounded and the number of t is infinite, (3) contradicts (5) for a suitable T. It follows that $\varphi^*(t) < 0$ for all real $t \neq 0$, that is, this quadratic form is negative definite. We have thus obtained a necessary condition for the convergence of the series ρ in (2).

Next, assume that, conversely, $\varphi^*(t)$ is negative definite. Then the function $\varphi^*(t)$ considered on the p-dimensional unit sphere $t't = 1$ has a maximum $-\mu < 0$, and, in view of the homogeneity, we have quite generally,

$$\varphi^*(t) \leqslant -\mu t't$$

for all real t, and so, in particular, for integral values of t. Since $\psi^*(t)$ is linear,

$$\psi^*(t) < \frac{\mu}{2} t't \tag{6}$$

for almost all integral t and, consequently, for a suitable choice of a constant C,

$$|a(t)| < Ce^{-(\mu/2)t't} \leqslant Ce^{(-\mu/2)(|t_1|+\cdots+|t_p|)}$$

for all integral t. This implies directly the absolute convergence of the series ρ. If we now assume that the coefficients of $\psi(t)$ depend on some parameters, but their absolute values all lie below a given bound, then the estimate (6) remains valid for almost all integral values of t. This implies the uniform convergence of the series in (2) relative to the coefficients of $\psi(t)$. We shall use this result first in the case $p = 1$, when it is almost trivial, and then in the general case.

By the Weierstrass product theorem, every meromorphic function and, in particular, every elliptic function $f(s)$ of a complex variable s can be written as a quotient

$$f(s) = \frac{g(s)}{h(s)} \tag{7}$$

of entire functions $g(s)$ and $h(s)$. We may also assume that the zeros of $g(s)$ and $h(s)$ yield precisely the zeros and poles of $f(s)$. Since $g(s)$ and $h(s)$ have no zeros in common, we shall call them *relatively prime*. For every period ω of $f(s)$ the function $h(s + \omega)$ has precisely the same zeros as $h(s)$. Hence the quotient $h(s + \omega)/h(s)$ is an entire function without zeros whose logarithm, after the choice of an as yet undetermined multiple of $2\pi i$, is single-valued and regular in the whole s-plane. It follows that there is an entire function $q(\omega, s)$ such that

$$h(s + \omega) = e^{q(\omega,s)}h(s), \qquad g(s + \omega) = e^{q(\omega,s)}g(s). \tag{8}$$

We shall show that if the elliptic function $f(s)$ is nonconstant, then the exponent $q(\omega, s)$ cannot be constant in s for every choice of ω. Otherwise, the first equality in (8) would imply that the logarithmic derivative

$$\frac{h'(s)}{h(s)} = l(s)$$

is itself an elliptic function of s for which all the ω are periods. Since $f(s)$ is not constant and there is no nonconstant entire elliptic function, $h(s)$ has zeros. Then the poles of $l(s)$ are precisely the zeros of $h(s)$, and the corresponding residues are the multiplicities of the respective zeros and, therefore, all positive. This contradicts Theorem 1 in Section 14 of Chapter 1 (Volume I) which asserts that the sum of the residues of an elliptic function is 0.

We select a basis ω_1, ω_2 of the period lattice and put $q(\omega_1, s) = q_1(s)$, $q(\omega_2, s) = q_2(s)$. We then have the two functional equations

$$h(s + \omega_1) = e^{q_1(s)}h(s), \qquad h(s + \omega_2) = e^{q_2(s)}h(s),$$

which make it a simple matter to express the function $q(\omega, s)$ explicitly in terms of the functions $q_1(s)$ and $q_2(s)$ for any period $\omega = n_1\omega_1 + n_2\omega_2$. In particular, we see by induction that $q(\omega, s)$ is a polynomial in s of degree not higher than m, provided that this is the case for $\omega = \omega_1$ and $\omega = \omega_2$. We shall show in Section 11 that in the quotient representation (7) of an arbitrary elliptic function $f(s)$, we can restrict ourselves to relatively prime entire functions $g(s)$ and $h(s)$ for which

$$q_1(s) = 0, \qquad q_2(s) = \alpha s + \beta \qquad (\alpha \neq 0)$$

and α, β are independent of s. Then, in all cases, $q(\omega, s)$ is linear in s. By replacing $\omega_1^{-1}s$ by s, we may suppose the fundamental periods to have the normalized form

$$\omega_1 = 1, \qquad \omega_2 = z = x + iy, \qquad y > 0.$$

Furthermore, if s is replaced by $s + c$ with constant c, then β goes over into $\beta + \alpha c$; so that, by assigning a suitable value to c, we can make β take on any desired value. For what follows it is convenient to introduce the normalization $\beta = (\alpha/2)z$ and, for complex a, the notation

$$e^{\pi i a} = \varepsilon(a).$$

Theorem 1: The two simultaneous functional equations

$$(9) \qquad h(s + 1) = h(s), \qquad h(s + z) = e^{\alpha(s+z/2)}h(s) \qquad (\alpha \neq 0)$$

have a solution $h(s)$ which is an entire function not identically zero, if and only if $\alpha = -2\pi im$ with m a natural number. In that case all the solutions are given by

$$(10) \quad h(s) = \sum_{r=1}^{m} b_r h_r(s), \qquad h_r(s) = \sum_{t \equiv r (\mathrm{mod}\ m)} \varepsilon\left(\frac{zt^2}{m} + 2ts\right) \qquad (r = 1, \ldots, m),$$

where $b_1, \ldots, b_m$ are arbitrary constants and t varies over the integers in the residue class of $r(\mathrm{mod}\ m)$.

Proof: To begin with we show that the series $h_r(s)$ represents an entire function of s. In fact, if we replace t by $mt + r$, then the general term of the series takes the form (2) with

$$\varphi(t) = \pi imzt^2, \qquad \psi(t) = 2\pi i(ms + rz)t,$$

$$c = \pi i \frac{z}{m} r^2 + 2rs, \qquad \varphi^*(t) = -\pi myt^2.$$

Hence $\varphi^*(t)$ is negative definite, and the coefficient $2\pi i(ms + rz)$ which appears in $\psi(t)$ is bounded when s varies over a bounded region $\mathfrak{G}$ of the s-plane. Since, in that case, the series converges uniformly in $\mathfrak{G}$ and the general term is an entire function of s, it follows, by the theorem of Weierstrass, that the sum $h_r(s)$ is regular in $\mathfrak{G}$. Since $\mathfrak{G}$ can be chosen arbitrarily, $h_r(s)$ is entire, as asserted.

Now let $h(s)$ be a nontrivial entire solution of (9). Then it has the period 1. The substitution

$$e^{2\pi is} = \varepsilon(2s) = v$$

makes $h(s)$ into a function of v which is single-valued and regular for all finite nonzero values of v, and has a Laurent expansion

$$(11) \qquad h(s) = \sum_{t=-\infty}^{\infty} a_t v^t = \sum_{t=-\infty}^{\infty} a_t \varepsilon(2ts)$$

with fully determined coefficients a_t. Since the function $h(s + z)$ also has the period 1 and $h(s)$ is not identically zero, the second equation in (9) implies the condition

$$e^{\alpha(s+1+z/2)} = e^{\alpha(s+z/2)} \qquad (\alpha \neq 0),$$

so that $\alpha = -2\pi i m$ with nonzero integer m. If, as in the proof of Theorem 2 in Section 14 of Chapter 1 (Volume I), we integrate the logarithmic derivative of $h(s)$ over the boundary of the parallelogram $\mathfrak{P}$ on the vectors 1 and z, then m is seen to be equal to the number of zeros of $h(s)$ on $\mathfrak{P}$, so that $m \geqslant 0$; hence $m > 0$. Later we shall deduce this inequality directly from the convergence of (11). Equation (11) and the second functional equation (9) imply

$$\sum_{t=-\infty}^{\infty} a_t \varepsilon(2tz) v^t = \sum_{t=-\infty}^{\infty} a_t \varepsilon(2ts - 2ms - mz) = \sum_{t=-\infty}^{\infty} a_{t+m} \varepsilon(-mz) v^t,$$

and, therefore, the recursion formula

$$a_t \varepsilon(2tz) = a_{t+m} \varepsilon(-mz).$$

If we put

$$a_t \varepsilon\left(\frac{-zt^2}{m}\right) = b_t,$$

then

$$b_{t+m} = a_t \varepsilon\left((m + 2t)z - \frac{(m + t)^2 z}{m}\right) = b_t.$$

From this we obtain

$$(12) \qquad a_t = b_r \varepsilon\left(\frac{zt^2}{m}\right),$$

provided that

$$t \equiv r (\operatorname{mod} |m|) \qquad (r = 1, 2, \ldots, |m|).$$

(11) and (12) imply the assertion in (10). In view of the convergence of (11) each of the series $b_r h_r(s)$ converges and, in addition, at least one coefficient $b_r \neq 0$. It follows that $\varphi^*(t)$ is negative definite, and so m is actually positive. At the same time the computation shows that each of the functions $h_r(s)$, and, therefore, also the sum $h(s)$ in (10) for any choice of the b_r, satisfies both of the functional equations (9). This completes the proof.

In the sequel we shall require only the special case $m = 1$ of Theorem 1. Accordingly, the equations

$$h(s + 1) = h(s), \qquad h(s + z) = \varepsilon(-2s - z)h(s)$$

have the entire solution

$$\Theta(s) = \sum_{t=-\infty}^{\infty} \varepsilon(t^2 z + 2ts) = \sum_{t=-\infty}^{\infty} e^{\pi i t^2 z + 2\pi i t s} \tag{13}$$

which is unique up to a factor independent of s. This is the *theta function* introduced by Jacobi, who based all of the theory of elliptic functions on its properties. The theta function also depends on the quantity z in the upper half plane so that we could, instead of $\Theta(s)$, write more accurately $\Theta(z, s)$. The more general functions appearing in Theorem 1 can easily be linked to $\Theta(z, s)$; in particular, we have the obvious relation

$$h_r(s) = \sum_{t=-\infty}^{\infty} \varepsilon\left(\frac{z(mt + r)^2}{m} + 2(mt + r)s\right) = \varepsilon\left(\frac{zr^2}{m} + 2rs\right)\Theta(mz, ms + rz).$$

In the eighteenth century the theta function appeared in a different connection in the work of Euler and, still earlier, in the work of Johann Bernoulli. Before Jacobi, Gauss recognized the importance of the function defined by (13) for the theory of elliptic functions, and developed its essential properties without, however, publishing these important discoveries. We are about to introduce the *theta function of p variables* $s_1, \ldots, s_p$. After the fundamental investigations of Riemann and Weierstrass, the theory of this function was decisively advanced by Frobenius and Poincaré. The creation of this theory may be safely regarded as one of the most significant accomplishments of the nineteenth century.

In the next chapter (Volume III) we shall concern ourselves in great detail with establishing the theory of functions of p complex variables $s_1, \ldots, s_p$. Subject to later explanations it will serve our present purpose to define an *entire function of s* as the sum of a power series

$$h(s) = \sum_{n_1, \ldots, n_p = 0}^{\infty} c_{n_1 \cdots n_p} s_1^{n_1} \cdots s_p^{n_p}$$

with constant complex coefficients $c_{n_1 \cdots n_p}$ supposed absolutely convergent for all finite values $s_1, \ldots, s_p$. With a view to generalizing the function $\Theta(z, s)$, which corresponds to the case $p = 1$, to the case $p \geqslant 1$, we consider a $p \times p$ symmetric complex matrix $Z = X + iY$ with imaginary part $Y > 0$, that is, Y positive definite. We again denote a column of integers $t_1, \ldots, t_p$ by t and put

$$\varphi(t) = \pi i t' Z t, \qquad \psi(t) = 2\pi i t' s, \qquad \varphi^*(t) = -\pi t' Y t,$$

$$\Theta(s) = \Theta(Z, s) = \sum_t \varepsilon(t' Z t + 2t' s) = \sum_t e^{\pi i t' Z t + 2\pi i t' s}. \tag{14}$$

It is clear that, for $p = 1$, (14) reduces to (13). Since $\varphi^*(t)$ is negative definite, the series in (14) converges absolutely. Also, the convergence is uniform in s if s is restricted to a bounded region of the complex s-space. By applying the Cauchy integral formula p times to the general term of (14) and interchanging the order of summation and integration we see that $\Theta(s)$ is actually representable by an invariably convergent power series in p variables $s_1, \ldots, s_p$, and so is an entire function of s. This is the *theta function in* p variables, which, for $p = 1$, is also referred to as the *elliptic theta function*. As t varies over all the columns of integers so does $-t$. It follows, from (14), that

$$\Theta(-s) = \Theta(s), \tag{15}$$

and it is in this sense that $\Theta(s)$ is an even function.

Theorem 2: The function

$$h(s) = \Theta(Z, s)$$

satisfies the simultaneous functional equations

$$h(s + g) = h(s), \qquad h(s + Zg) = \varepsilon(-g'Zg - 2g's)h(s) \tag{16}$$

for all integer columns g; and, every other solution which is an entire function is obtained from it by multiplication by a constant independent of s.

Proof: The proof of Theorem 1 can be carried over without difficulty. If an entire function satisfies the first condition (16) for all g, then it has the period 1 in each variable, and so has the Laurent expansion

$$h(s) = \sum_t a_t \varepsilon(2t's)$$

with uniquely determined coefficients a_t. The second condition implies further

$$\sum_t a_t \varepsilon(2t'Zg)\varepsilon(2t's) = \sum_t a_t \varepsilon(-g'Zg - 2g's + 2t's) = \sum_t a_{t+g}\varepsilon(-g'Zg)\varepsilon(2t's),$$

so that

$$a_t \varepsilon(2t'Zg) = a_{t+g}\varepsilon(-g'Zg).$$

If we put

$$a_t \varepsilon(-t'Zt) = b_t,$$

then we obtain

$$b_{t+g} = a_t \varepsilon(2t'Zg + g'Zg - (t + g)'Z(t + g)) = b_t.$$

Our assertion follows, since b_0 can be chosen as an arbitrary magnitude independent of s, and, at the same time, the computation shows that then $h(s)$ actually has the required properties.

10. The zeros of the theta function

When we defined $\Theta(Z, s)$ by (14) in the preceding section, we supposed Z a $p \times p$ symmetric complex matrix with positive imaginary part Y. It follows that the $p(p+1)/2$ elements z_{kl} $(1 \leqslant k \leqslant l \leqslant p)$ of Z may be regarded as independent complex variables subject only to the condition that Z lies in the generalized upper half plane. For the purposes of this chapter, however, we suppose Z a definite fixed period matrix formed from a normal basis of integrals of the first kind and a definite canonical dissection of the given Riemann surface $\mathfrak{R}$ of genus $p > 0$. Viewed as a function of the p independent variables $s_1, \ldots, s_p$, $\Theta(s)$ is not identically zero since all of the coefficients in (14), its Laurent expansion in powers of $v_k = \varepsilon(2s_k)$ $(k = 1, \ldots, p)$, are $\neq 0$. Riemann perceived the intimate connection between the zeros of $\Theta(s)$ and the explicit solution of the inversion problem. His results will be discussed in this and the next section.

Let $\mathfrak{p}$ be a variable point of $\mathfrak{R}$ and $w(\mathfrak{p})$ the column formed with the normal basis $w_1, \ldots, w_p$. Let A_k, B_k $(k = 1, \ldots, p)$ be the pairs of conjugate crosscuts with common point $\mathfrak{p}_0$ which we also regard as the initial point of the boundary arc A_1 on the dissected Riemann surface $\mathfrak{R}^*$. The point $\mathfrak{p}_0$ is also the initial point for integration, so that $w(\mathfrak{p})$ is uniquely determined on $\mathfrak{R}^*$ by the condition $w(\mathfrak{p}_0) = 0$. It was Riemann's fundamental idea to introduce the function

$$\varphi(\mathfrak{p}) = \varphi(\mathfrak{p}, u) = \Theta(w(\mathfrak{p}) - u) \tag{1}$$

with variable $\mathfrak{p}$ and constant u. Since $w(\mathfrak{p})$ is regular on $\mathfrak{R}^*$ and, for variable s, $\Theta(s)$ is an entire function, $\varphi(\mathfrak{p})$ is regular on $\mathfrak{R}^*$. Furthermore, it is clear that $\varphi(\mathfrak{p})$ can be continued analytically and is regular along every path issuing from $\mathfrak{p}_0$ on the uncut surface $\mathfrak{R}$ without, however, being single-valued in the large there due to the periods of $w(\mathfrak{p})$ which enter with every circuit. With every circuit of $\mathfrak{p}$ along a closed curve on $\mathfrak{R}$, $w(\mathfrak{p})$ changes by a summand of the form $g + Zh$ with integer columns g and h. At the same time, by Theorem 2 of the preceding section, $\varphi(\mathfrak{p})$ is multiplied by the nonzero factor $\varepsilon(-h'Zh - 2h's)$ with $s = w(\mathfrak{p}) - u$, and this leads to the multiple-valuedness of $\varphi(\mathfrak{p})$ on $\mathfrak{R}$. In particular, this shows that $\varphi(\mathfrak{p})$ has the same set of zeros on $\mathfrak{R}$ and on $\mathfrak{R}^*$. Since $\Theta(s)$ does not vanish identically in s, it follows that, in any case, there exist constant u such that $\varphi(\mathfrak{p})$, viewed as a function of the variable $\mathfrak{p}$ alone, is not identically zero.

Theorem 1: If, for a fixed u, the function $\varphi(\mathfrak{p})$ does not vanish identically in $\mathfrak{p}$, then it has exactly p zeros on $\mathfrak{R}$.

Proof: It is possible to modify the canonical dissection by a suitable small deformation so that no zero of the nontrivial function $\varphi(\mathfrak{p})$ is on the

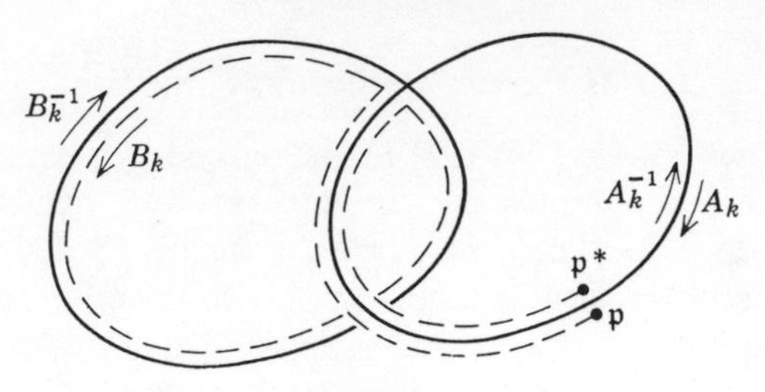

Figure 44

boundary C of $\mathfrak{R}^*$. The required number n of zeros is equal to the change in the function $(1/2\pi i) \log \varphi(\mathfrak{p})$ when C is traversed in the positive direction. The boundary C is the result of joining the $4p$ paths A_k, B_k, A_k^{-1}, B_k^{-1} ($k = 1, \ldots, p$) in this order with $\mathfrak{p}_0$, the initial point of A_1 on $\mathfrak{R}^*$. When A_k is traversed, then $w(\mathfrak{p})$ is augmented by the summand $w(A_k)$; and, similarly, when B_k is traversed, then $w(\mathfrak{p})$ is augmented by the summand $w(B_k)$. The matrix formed with the $2p$ columns $w(A_1), \ldots, w(A_p), w(B_1), \ldots, w(B_p)$ is precisely the transposed period matrix (EZ). Now we argue in a manner analogous to that of the proof of Theorem 1 in Section 3. Specifically, let $\mathfrak{p}$ and $\mathfrak{p}^*$ be two points on the boundary arcs A_k and A_k^{-1} of $\mathfrak{R}^*$ which coincide on $\mathfrak{R}$ (Figure 44). If we go on $\mathfrak{R}^*$ from $\mathfrak{p}$ along any path to $\mathfrak{p}^*$, then $\varphi(\mathfrak{p})$ is multiplied by the factor

$$\varepsilon(-e_k'Ze_k - 2e_k'(w(\mathfrak{p}) - u)) = \varepsilon(-z_{kk} + 2u_k - 2w_k(\mathfrak{p})),$$

where e_k is the kth column of the $p \times p$ unit matrix, and we have

$$\varphi(\mathfrak{p}^*) = \varepsilon(-z_{kk} + 2u_k - 2w_k(\mathfrak{p}))\varphi(\mathfrak{p}) \qquad (\mathfrak{p} \text{ on } A_k). \tag{2}$$

On the other hand, if $\mathfrak{p}$ and $\mathfrak{p}^*$ are two points on the boundary arcs B_k and B_k^{-1} which coincide on $\mathfrak{R}$ (Figure 45), and we go on $\mathfrak{R}^*$, along any path, from $\mathfrak{p}$ to $\mathfrak{p}^*$, then $w(\mathfrak{p})$ is augmented by the summand $w(A_k^{-1}) = -e_k$,

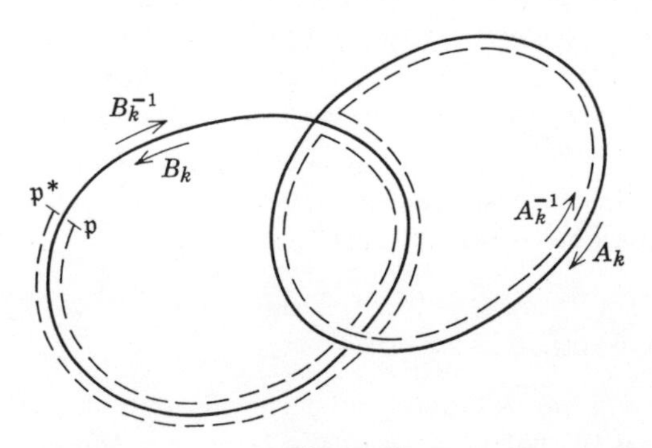

Figure 45

but $\varphi(\mathfrak{p})$ remains unchanged, and we have

$$\varphi(\mathfrak{p}^*) = \varphi(\mathfrak{p}) \qquad (\mathfrak{p} \text{ on } B_k). \tag{3}$$

Now we note that the boundary arcs A_k and A_k^{-1} as well as B_k and B_k^{-1} of $\mathfrak{R}^*$ have opposite orientations on $\mathfrak{R}$. The contribution of the four successive boundary arcs A_k, B_k, A_k^{-1}, B_k^{-1} to the total change in the function $(1/2\pi i) \log \varphi(\mathfrak{p})$ upon traversal of C is, by (2) and (3), equal to the change in $\frac{1}{2}(2u_k - z_{kk} - 2w_k(\mathfrak{p}))$ upon traversal of A_k^{-1}, that is, 1. Summing over k we see that the required total change is p, so that $n = p$, as asserted.

If u is chosen so that the function $\varphi(\mathfrak{p}, u) = \varphi(\mathfrak{p})$ does not vanish identically in $\mathfrak{p}$, then, by Theorem 1, the p zeros $\mathfrak{q}_1, \ldots, \mathfrak{q}_p$ of $\varphi(\mathfrak{p})$ furnish an integral divisor $\mathfrak{q} = \mathfrak{q}_1 \cdots \mathfrak{q}_p$ of degree p. It turns out that it is this divisor which yields the solution of the inversion problem. Specifically, let c be the column consisting of the p quantities

$$c_k = -\tfrac{1}{2}z_{kk} + \sum_{l=1}^{p} \int_{A} w_k \, dw \qquad (k = 1, \ldots, p), \tag{4}$$

where the integration extends over the boundary arc A_l of $\mathfrak{R}^*$, and let

$$s = u + c. \tag{5}$$

Then we have

Theorem 2: If, for a fixed s, the function

$$\varphi(\mathfrak{p}) = \Theta(w(\mathfrak{p}) - s + c)$$

does not vanish identically in $\mathfrak{p}$, then its p zeros $\mathfrak{q}_1, \ldots, \mathfrak{q}_p$ satisfy the congruence

$$\sum_{k=1}^{p} w(\mathfrak{q}_k) \equiv s. \tag{6}$$

Proof: In analogy to the proof of Theorem 1 we integrate the expression $w(\mathfrak{p}) \, d \log \varphi(\mathfrak{p})$ over the boundary C of $\mathfrak{R}^*$. By the residue theorem,

$$\frac{1}{2\pi i} \int_C w \, d \log \varphi = \sum_{k=1}^{p} w(\mathfrak{q}_k) \tag{7}$$

is precisely the sum in (6). If $\mathfrak{p}$ and $\mathfrak{p}^*$ have the same meaning as in the earlier proof, then (2) and (3) yield the formulas

$$d \log \varphi(\mathfrak{p}^*) = -2\pi i \, dw_k(\mathfrak{p}) + d \log \varphi(\mathfrak{p}) \qquad (\mathfrak{p} \text{ on } A_k),$$

$$d \log \varphi(\mathfrak{p}^*) = d \log \varphi(\mathfrak{p}) \qquad (\mathfrak{p} \text{ on } B_k);$$

while, on the other hand,

$$w(\mathfrak{p}^*) = w(\mathfrak{p}) + w(B_k) \qquad (\mathfrak{p} \text{ on } A_k),$$

$$w(\mathfrak{p}^*) = w(\mathfrak{p}) + w(A_k^{-1}) \qquad (\mathfrak{p} \text{ on } B_k).$$

We thus obtain

$$\int_C w\, d\log\varphi = \sum_{l=1}^{p}\left(\int_{A_l} + \int_{B_l} + \int_{A_l^{-1}} + \int_{B_l^{-1}}\right), \tag{8}$$

$$\frac{1}{2\pi i}\left(\int_{A_l} + \int_{A_l^{-1}}\right) = \frac{1}{2\pi i}\int_{A_l}\{w(\mathfrak{p})\, d\log\varphi(\mathfrak{p}) - (w(\mathfrak{p}) + w(B_l))\cdot \tag{9}$$

$$\cdot(-2\pi i\, dw_l(\mathfrak{p}) + d\log\varphi(\mathfrak{p}))\}$$

$$= w(B_l)w_l(A_l) + \int_{A_l} w\, dw_l - \frac{1}{2\pi i}\, w(B_l)\int_{A_l} d\log\varphi,$$

$$\frac{1}{2\pi i}\left(\int_{B_l} + \int_{B_l^{-1}}\right) = \frac{1}{2\pi i}\int_{B_l}\{w(\mathfrak{p})\, d\log\varphi(\mathfrak{p}) \tag{10}$$

$$- (w(\mathfrak{p}) - w(A_l))\, d\log\varphi(\mathfrak{p})\}$$

$$= \frac{1}{2\pi i}\, w(A_l)\int_{B_l} d\log\varphi.$$

By (3), the change in the value of $(1/2\pi i)\log\varphi(\mathfrak{p})$ over A_l is an integer. On the other hand, by (2), the corresponding change over B_l is equal to $\frac{1}{2}(-z_{ll} + 2u_l - 2w_l(A_l))$ plus an integer, where we point out that on $\mathfrak{R}^*$, the value of w_l at the terminal point of A_l is actually $w_l(A_l)$. Since $w(A_l) = e_l$, (8), (9), (10) yield the relation

$$\frac{1}{2\pi i}\int_C w\, d\log\varphi \equiv \sum_{l=1}^{p}\left\{e_l(u_l - \tfrac{1}{2}z_{ll}) + \int_{A_l} w\, dw_l\right\},$$

and so, by (4), (5), (7), the assertion (6).

In consequence of Theorem 2

$$\sum_{k=1}^{p}\int_{\mathfrak{p}_0}^{\mathfrak{q}_k} dw \equiv s,$$

and, quite generally,

$$\sum_{k=1}^{p}\int_{\mathfrak{a}_k}^{\mathfrak{q}_k} dw \equiv s + \sum_{k=1}^{p}\int_{\mathfrak{a}_k}^{\mathfrak{p}_0} dw,$$

where the integration extends along arbitrary paths on $\mathfrak{R}$ connecting the indicated limits. If $\varphi(\mathfrak{p})$ does not vanish identically in $\mathfrak{p}$, then its p zeros yield the solution of the Jacobi inversion problem. The remaining theorems in this section deal with the special case when $\varphi(\mathfrak{p})$ vanishes identically.

Theorem 3: For

$$\Theta(t - c) = 0 \tag{11}$$

to hold it is necessary and sufficient that the congruence

$$t \equiv \sum_{l=2}^{p} w(\mathfrak{y}_l) \tag{12}$$

is satisfied for $p - 1$ suitably chosen points $\mathfrak{y}_2, \ldots, \mathfrak{y}_p$.

Proof: Let $\mathfrak{z}_1, \ldots, \mathfrak{z}_p$ be p distinct points on $\mathfrak{R}$. We put

$$s = \sum_{k=1}^{p} w(\mathfrak{z}_k). \tag{13}$$

The integral divisor $\mathfrak{z} = \mathfrak{z}_1 \cdots \mathfrak{z}_p$ of degree p is general provided that the determinant $|dw_l/d\mathfrak{z}_k| \neq 0$; then the transformation given by (13) is locally uniquely invertible. Since the function $\Theta(w(\mathfrak{p}) - s + c)$ does not vanish identically in s, it is possible, given an integral divisor $\mathfrak{y} = \mathfrak{y}_1 \cdots \mathfrak{y}_p$ of degree p, to find a sequence $\mathfrak{z}_k$ for $k = 1, \ldots, p$ with $\mathfrak{z}_k \to \mathfrak{y}_k$ such that $\mathfrak{z}$ is general throughout and $\Theta(w(\mathfrak{p}) - s + c)$ does not vanish identically in $\mathfrak{p}$ when s is determined by (13). Then, in view of Theorem 2, the function $\varphi(\mathfrak{p})$ has altogether p zeros $\mathfrak{q}_1, \ldots, \mathfrak{q}_p$ on $\mathfrak{R}$ which satisfy the congruence (6). Since $\mathfrak{z}$ is general, we conclude with the aid of (13) that, after proper indexing, $\mathfrak{q}_k = \mathfrak{z}_k$ $(k = 1, \ldots, p)$. In particular,

$$\Theta(w(\mathfrak{z}_1) - s + c) = 0.$$

By (15) in Section 9 and (13) we also have

$$\Theta\left(-c + \sum_{k=2}^{p} w(\mathfrak{z}_k)\right) = 0.$$

Upon transition to the limit $\mathfrak{z}_k \to \mathfrak{y}_k$ $(k = 2, \ldots, p)$, we find that the condition (12) is sufficient for (11) to hold.

Conversely, let (11) hold for a column t. Then there exists a smallest natural number $m \leqslant p$ such that, with $\mathfrak{x}_1, \ldots, \mathfrak{x}_m$ and $\mathfrak{y}_1, \ldots, \mathfrak{y}_m$ varying independently, the function

$$\varphi(\mathfrak{x}_1) = \Theta(w(\mathfrak{x}_1) - s + c)$$

with

$$s = t - \sum_{k=2}^{m} w(\mathfrak{x}_k) + \sum_{k=1}^{m} w(\mathfrak{y}_k) \tag{14}$$

does not vanish identically in all the $2m$ variables. Now we suppose the $2m - 1$ points $\mathfrak{x}_2, \ldots, \mathfrak{x}_m$ and $\mathfrak{y}_1, \ldots, \mathfrak{y}_m$ determined so that $\varphi(\mathfrak{x}_1)$ does not vanish identically in $\mathfrak{x}_1$. Then, by Theorems 1 and 2, the function $\varphi(\mathfrak{x}_1)$

has exactly p zeros $\mathfrak{q}_1, \ldots, \mathfrak{q}_p$ on $\mathfrak{R}$ for which (6) holds. On the other hand, in view of the minimum property of m we also have

$$\varphi(\mathfrak{y}_k) = 0 \qquad (k = 1, \ldots, m).$$

If $\mathfrak{y}_1, \ldots, \mathfrak{y}_m$ are chosen to be distinct, then with proper indexing we must have $\mathfrak{y}_k = \mathfrak{q}_k$ $(k = 1, \ldots, m)$. But then, by (6) and (14),

$$t \equiv \sum_{k=2}^{m} w(\mathfrak{x}_k) + \sum_{k=m+1}^{p} w(\mathfrak{q}_k).$$

This proves the assertion in (12).

Theorem 3 yields a representation of the totality of zeros of $\Theta(s)$ using $p - 1$ independent parameters. If $\varphi(\mathfrak{p})$ vanishes identically in $\mathfrak{p}$, then (15) in Section 9 and Theorem 3 imply

$$s \equiv w(\mathfrak{p}) + \sum_{l=2}^{p} w(\mathfrak{y}_l)$$

with $\mathfrak{p}$ arbitrary and $\mathfrak{y}_2, \ldots, \mathfrak{y}_p$ chosen appropriately. Then, using the notation introduced at the end of Section 8, s is special. In order to complete and improve Theorem 2, we investigate analogously the inversion problem with no restrictions on s.

Theorem 4: Given s, let $m + 1$ be the smallest natural number such that the function

$$\varphi_m(\mathfrak{p}) = \Theta\left(w(\mathfrak{p}) - s + c + \sum_{k=1}^{m}\{w(\mathfrak{y}_k) - w(\mathfrak{x}_k)\}\right)$$

does not vanish identically in $\mathfrak{p}$ and the $2m$ parameters $\mathfrak{x}_k, \mathfrak{y}_k$ $(k = 1, \ldots, m)$. Then s has rank $p - m$. Furthermore, if $\mathfrak{x}_k$ are m distinct points and $\mathfrak{y}_k$ are m additional points chosen so that $\varphi_m(\mathfrak{p})$ does not vanish identically in $\mathfrak{p}$, then the zeros of this function are $\mathfrak{x}_1, \ldots, \mathfrak{x}_m$ and $p - m$ points $\mathfrak{q}_{m+1}, \ldots, \mathfrak{q}_p$, and we have

$$s \equiv \sum_{k=1}^{m} w(\mathfrak{y}_k) + \sum_{k=m+1}^{p} w(\mathfrak{q}_k). \tag{15}$$

Proof: Let the $2m$ points $\mathfrak{x}_k, \mathfrak{y}_k$ $(k = 1, \ldots, m)$ satisfy the assumptions above. If we put

$$v = s + \sum_{k=1}^{m}\{w(\mathfrak{x}_k) - w(\mathfrak{y}_k)\}, \tag{16}$$

then we can apply Theorem 2 with v in place of s and obtain for the p zeros $\mathfrak{q}_1, \ldots, \mathfrak{q}_p$ of the function $\varphi_m(\mathfrak{p})$ the congruence

$$\sum_{k=1}^{p} w(\mathfrak{q}_k) \equiv v. \tag{17}$$

In view of the minimum property of m, the m distinct points $\mathfrak{x}_1, \ldots, \mathfrak{x}_m$

are all zeros of $\varphi_m(\mathfrak{p})$, so that we can put $\mathfrak{x}_k = \mathfrak{q}_k$ $(k = 1, \ldots, m)$ and obtain (15) from (16) and (17).

To begin with, we impose on the points $\mathfrak{y}_1, \ldots, \mathfrak{y}_m$ the condition that $\varphi_m(\mathfrak{p})$ does not vanish identically in $\mathfrak{p}$ and $\mathfrak{x}_1, \ldots, \mathfrak{x}_m$. By going over to the limit we see that (15) holds for arbitrarily prescribed $\mathfrak{y}_1, \ldots, \mathfrak{y}_m$ and suitably chosen $\mathfrak{q}_{m+1}, \ldots, \mathfrak{q}_p$. It follows that the rank of the divisor $\mathfrak{y}_1 \cdots \mathfrak{y}_m \mathfrak{q}_{m+1} \cdots \mathfrak{q}_p$, and so the rank of s, is a number $r \leqslant p - m$. If we had $r < p - m$, then, in (15), we could also prescribe $\mathfrak{q}_p = \mathfrak{p}$ arbitrarily. If we put

$$t = s - w(\mathfrak{p}) + \sum_{k=1}^{m} \{w(\mathfrak{x}_k) - w(\mathfrak{y}_k)\},$$

then

$$\varphi_m(\mathfrak{p}) = \Theta(c - t) = \Theta(t - c)$$

and

$$t \equiv \sum_{k=1}^{m} w(\mathfrak{x}_k) + \sum_{k=m+1}^{p-1} w(\mathfrak{q}_k).$$

By Theorem 3, this would lead to a contradiction, for $\varphi_m(\mathfrak{p})$ would vanish identically in $\mathfrak{p}$ and the parameters $\mathfrak{x}_k, \mathfrak{y}_k$ $(k = 1, \ldots, m)$. It follows that $r = p - m$ and the proof is complete.

Theorem 4 implies, in particular, that s is general, and so the inversion problem is uniquely solvable, if and only if the function

$$\varphi(\mathfrak{p}) = \Theta(w(\mathfrak{p}) - s + c)$$

does not vanish identically in $\mathfrak{p}$. For $p = 1$ this condition is satisfied for every choice of s. Furthermore, in this case, Theorem 3 implies that the function $\Theta(s)$ of the one complex variable s vanishes precisely for $s \equiv -c$. If 1 and z are the fundamental periods of the normal integral w of the first kind associated with the given canonical dissection, then we obtain from (4) for $p = 1$ the formula

$$c = -\tfrac{1}{2}z + \tfrac{1}{2}(w(A_1))^2 = \frac{1 - z}{2}.$$

It follows that $s = (1 + z)/2$ is the only zero of the elliptic theta function $\Theta(z, s)$ in the period parallelogram. Since $w(\mathfrak{p})$ maps the canonically dissected Riemann surface conformally onto a fundamental region of the period group, it follows more precisely from Theorem 1 that the zero in question is simple. Incidentally, from (13) in Section 9 we obtain, by direct computation, that, in fact,

$$\begin{aligned}
\Theta\left(z, \frac{1+z}{2}\right) &= \sum_{t=-\infty}^{\infty} \varepsilon(t^2 z + t(1 + z)) \\
&= \sum_{t=0}^{\infty} \{\varepsilon(t^2 z + t(1 + z)) + \varepsilon((-t - 1)^2 z + (-t - 1)(1 + z))\} \\
&= \sum_{t=0}^{\infty} ((-1)^t + (-1)^{-t-1})\varepsilon(t^2 z + tz) = 0.
\end{aligned}$$

However, there is no correspondingly simple way of verifying the simplicity of this zero.

11. Theta quotients

To begin with we suppose $p = 1$. Then $\Theta(s)$ is an entire function of the complex variable s with one (simple) zero at $(1 + z)/2$ throughout the period parallelogram $\mathfrak{P}$ on 1 and z which satisfies the equations

$$\Theta(s + 1) = \Theta(s), \qquad \Theta(s + z) = \varepsilon(-2s - z)\Theta(s), \qquad \Theta(-s) = \Theta(s).$$

It follows that, with $c = (1 - z)/2$ and arbitrary a,

$$\Theta(u - a - c) = \theta(u, a) = \theta(u)$$

is an entire function of u which satisfies the equations

$$\theta(u + 1) = \theta(u), \qquad \theta(u + z) = -\varepsilon(2a - 2u - 2z)\theta(u), \tag{1}$$

and its only zero in $\mathfrak{P}$ is congruent to a. Now let $a_1, \ldots, a_m$ and $b_1, \ldots, b_m$ be $2m$ complex numbers such that

$$a_1 + \cdots + a_m = b_1 + \cdots + b_m \tag{2}$$

and

$$a_k \neq b_l \qquad (k, l = 1, \ldots, m). \tag{3}$$

Here the various a_k and the various b_k need not be distinct. By (2) and (3) $m \geqslant 2$. If we put

$$F(u) = \prod_{k=1}^{m} \frac{\theta(u, a_k)}{\theta(u, b_k)} = \prod_{k=1}^{m} \frac{\Theta(u - a_k - c)}{\Theta(u - b_k - c)}, \tag{4}$$

then $F(u)$ is a meromorphic function of u which, by (1) and (2), has the periods 1 and z. Furthermore, its zeros are precisely at all the points which are congruent to the a_k $(k = 1, \ldots, m)$ and its poles are precisely at all the points which are congruent to the b_k $(k = 1, \ldots, m)$. It follows that $F(u)$ is an elliptic function associated with the period lattice which has the above poles and zeros. Multiplying $F(u)$ by a nonzero constant, we obtain the most general elliptic function of this kind; and, in view of Theorems 2 and 4 in Section 14 of Chapter 1 (Volume I), we obtain in this way all the nonconstant elliptic functions. In view of the form of the right-hand side of (4), we say that every elliptic function is expressible as a *theta quotient*; here we are overlooking the constant factor. In this multiplicative representation of the elliptic functions the theta function $\Theta(u - a - c)$ plays a role analogous to that of the linear function $u - a$ vis-à-vis the rational functions of u. As perceived by Jacobi, this connection makes it possible to develop the whole theory of the elliptic functions starting with the properties of the theta function. We also note that there is a simple relation connecting the theta function and the function $\sigma(u)$ introduced by Weierstrass, with the second derivative of log $\sigma(u)$ leading to the $\wp$-function. We shall not go into this matter, however.

Let $f(\mathfrak{p})$ be a nonconstant function in an algebraic function field of genus $p = 1$ with zeros $\mathfrak{a}_1, \ldots, \mathfrak{a}_m$ and poles $\mathfrak{b}_1, \ldots, \mathfrak{b}_m$. By Abel's theorem

$$\sum_{k=1}^{m} w(\mathfrak{a}_k) \equiv \sum_{k=1}^{m} w(\mathfrak{b}_k), \tag{5}$$

where, as in the preceding section, $w(\mathfrak{p})$ denotes the normal integral of the first kind. The paths of integration may be supposed determined so that we have an equality sign in (5). Then, with

$$w(\mathfrak{a}_k) = a_k, \qquad w(\mathfrak{b}_k) = b_k \qquad (k = 1, \ldots, m),$$

the condition (2) is satisfied and, furthermore, (3) holds, because $w(\mathfrak{p})$ maps the canonically dissected Riemann surface onto a schlicht region. The substitution $w(\mathfrak{p}) = u$ makes $f(\mathfrak{p})$ into an elliptic function whose zeros lie precisely at all the points which are congruent to a_k ($k = 1, \ldots, m$) and whose poles lie precisely at all the points which are congruent to b_k ($k = 1, \ldots, m$). Hence

$$f(\mathfrak{p}) = \gamma \prod_{k=1}^{m} \frac{\Theta(w(\mathfrak{p}) - w(\mathfrak{a}_k) - c)}{\Theta(w(\mathfrak{p}) - w(\mathfrak{b}_k) - c)} \tag{6}$$

for a suitable constant γ. The determination of γ will not be discussed here. The problem now is to extend the formula (6) in a natural way to the case $p > 1$ and to express every nonconstant function of an arbitrary algebraic function field K of genus p as a theta quotient.

Theorem 3 of the preceding section suggests how the individual factors in (6) should be modified in the case $p > 1$. In that case we choose $p - 1$ points $\mathfrak{x}_2, \ldots, \mathfrak{x}_p$ on the Riemann surface $\mathfrak{R}$, and form, using an additional fixed point $\mathfrak{a}$ and a variable point $\mathfrak{p}$, the function

$$\varphi(\mathfrak{p}, \mathfrak{a}) = \Theta\left(w(\mathfrak{p}) - w(\mathfrak{a}) - c + \sum_{k=2}^{p} w(\mathfrak{x}_k)\right).$$

By the theorem just mentioned, $\varphi(\mathfrak{p}, \mathfrak{a})$ has, as in the case $p = 1$, the zero $\mathfrak{p} = \mathfrak{a}$. For $p > 1$, however, there arise now two difficulties. If $\mathfrak{x}_2, \ldots, \mathfrak{x}_p$ are improperly chosen, then the function $\varphi(\mathfrak{p}, \mathfrak{a})$ may vanish identically in $\mathfrak{p}$ and this is certainly the case for $\mathfrak{x}_p = \mathfrak{a}$. Nevertheless, for a given $\mathfrak{a}$ it is possible to choose $\mathfrak{x}_2, \ldots, \mathfrak{x}_p$ so that this does not happen. But now, by Theorem 1 of the preceding section, the function $\varphi(\mathfrak{p}, \mathfrak{a})$ has altogether p zeros on $\mathfrak{R}$ that is, in addition to the prescribed zero $\mathfrak{a}$, $p - 1$ unwanted zeros. We wish to show that these zeros drop out as a result of the formation of the theta quotients. To this end we embark on the following preliminary considerations.

Let $\mathfrak{x} = \mathfrak{x}_2 \cdots \mathfrak{x}_p$ be an integral divisor of degree $p - 1$. We assume that it is general which does not mean that $\mathfrak{x}_2, \ldots, \mathfrak{x}_p$ need all be distinct. Thus there is no constant function f in K with the property $\mathfrak{x}^{-1} \mid f$. By the Riemann–Roch theorem there is a differential dv, unique up to a multiplicative constant, such that $\mathfrak{x} \mid (dv)$. By Theorem 3, Section 6, dv has a total of $2p - 2$ zeros of which $p - 1$ are $\mathfrak{x}_2, \ldots, \mathfrak{x}_p$. If $\mathfrak{y}_2, \ldots, \mathfrak{y}_p$ are the remaining $p - 1$ zeros, then the integral divisor $\mathfrak{y} = \mathfrak{y}_2 \cdots \mathfrak{y}_p$ of degree $p - 1$ is uniquely determined by $\mathfrak{x}$, and we have

$$(dv) = \mathfrak{x}\mathfrak{y}. \tag{7}$$

We now show that, just like $\mathfrak{x}$, $\mathfrak{y}$ is general. In fact, if $\mathfrak{y}^{-1} \mid (f)$ for a function f in K, then the relation

$$(f\,dv) = \mathfrak{x}\mathfrak{y}(f)$$

implies that $f\,dv$ is also a differential of the first kind and $\mathfrak{x} \mid (f\,dv)$. Since (dv) is uniquely determined, f must be constant and the assertion concerning $\mathfrak{y}$ follows. We conclude that for every general integral divisor $\mathfrak{x}$, there is precisely one general integral divisor $\mathfrak{y}$ of the same degree such that an equation of the form (7) holds; consequently, $\mathfrak{x}\mathfrak{y}$ is the divisor of the zeros of a differential of the first kind. It is clear, conversely, that $\mathfrak{x}$ is uniquely determined by $\mathfrak{y}$. We call $\mathfrak{x}$ and $\mathfrak{y}$ *complementary*.

Now let $\mathfrak{x} = \mathfrak{x}_2 \cdots \mathfrak{x}_p$ be general, $\mathfrak{y} = \mathfrak{y}_2 \cdots \mathfrak{y}_p$ complementary to $\mathfrak{x}$ and $\mathfrak{x}_1$ an arbitrary point. We wish to show that the integral divisor $\mathfrak{z} = \mathfrak{x}_1\mathfrak{x}$ of degree p is general if and only if $\mathfrak{x}_1$ is different from the $p - 1$ points $\mathfrak{y}_2, \ldots, \mathfrak{y}_p$ uniquely determined by $\mathfrak{x}$. To begin with, it follows from the Riemann–Roch theorem that $\mathfrak{z}$ is general if and only if there is no differential dv of the first kind other than 0 such that $\mathfrak{z} \mid dv$. On the other hand, if the formula $\mathfrak{z} \mid dv$ holds for a nontrivial differential of the first kind then surely $\mathfrak{z} \mid (dv)$ holds, and, therefore, also (7). It follows that $\mathfrak{x}_1$ must be one of the points $\mathfrak{y}_2, \ldots, \mathfrak{y}_p$, while, conversely, the assumption concerning dv is satisfied by just such a choice of $\mathfrak{x}_1$. This again proves the assertion.

Throughout the rest of this section dv_0 is a fixed nontrivial differential of the first kind, and $\mathfrak{a} = \mathfrak{a}_1\mathfrak{a}_2 \cdots \mathfrak{a}_{p-2}$ is the divisor of the zeros of dv_0, so that $(dv_0) = \mathfrak{a}$. We put

$$\sum_{k=1}^{2p-2} w(\mathfrak{a}_k) = q.$$

If dv is any nontrivial differential of the first kind and $\mathfrak{b} = \mathfrak{b}_1\mathfrak{b}_2 \cdots \mathfrak{b}_{2p-2}$ is the divisor of its zeros, then the quotient $dv/dv_0 = f$ is a function in the field K, and

$$(f) = \frac{\mathfrak{b}}{\mathfrak{a}} \tag{8}$$

implies, by Abel's theorem, the congruence

$$\sum_{k=1}^{2p-2} w(\mathfrak{b}_k) \equiv q. \tag{9}$$

Conversely, if this congruence holds for the $2p - 2$ points $\mathfrak{b}_k$ $(k = 1, \ldots, 2p - 2)$, then (8) is satisfied with f a function in K, and the expression $f\,dv_0 = dv$ turns out to be a differential of the first kind with the prescribed zeros $\mathfrak{b}_k$. It follows that an integral divisor $\mathfrak{b} = \mathfrak{b}_1\mathfrak{b}_2 \cdots \mathfrak{b}_{2p-2}$ of degree $2p - 2$ is the divisor of the zeros of a differential of the first kind if and only if the condition (9) holds.

Theorem 1: Let $\mathfrak{x}_2 \cdots \mathfrak{x}_p$ be a general divisor, and let $\mathfrak{y}_2 \cdots \mathfrak{y}_p$ be the divisor complementary to it. If the point $\mathfrak{a}$ is different from $\mathfrak{x}_2, \ldots, \mathfrak{x}_p$, then the function

$$\varphi(\mathfrak{p}, \mathfrak{a}) = \Theta\left(w(\mathfrak{p}) - w(\mathfrak{a}) - c + \sum_{k=2}^{p} w(\mathfrak{x}_k)\right)$$

has on $\mathfrak{R}$ precisely the p zeros $\mathfrak{a}, \mathfrak{y}_2, \ldots, \mathfrak{y}_p$. Furthermore, $q \equiv 2c$.

Proof: By Theorem 3 in the preceding section, a point $\mathfrak{p} = \mathfrak{z}$ is a zero of $\varphi(\mathfrak{p}, \mathfrak{a})$ if and only if, for a suitable choice of $p - 1$ points $\mathfrak{r}_2, \ldots, \mathfrak{r}_p$, we have the congruence

$$w(\mathfrak{z}) - w(\mathfrak{a}) + \sum_{k=2}^{p} w(\mathfrak{x}_k) \equiv \sum_{k=2}^{p} w(\mathfrak{r}_k).$$

By Abel's theorem this congruence is equivalent to the relation

$$\mathfrak{z}\mathfrak{x}_2 \cdots \mathfrak{x}_p \sim \mathfrak{a}\mathfrak{r}_2 \cdots \mathfrak{r}_p. \tag{10}$$

On the other hand, the divisor $\mathfrak{z}\mathfrak{x}_2 \cdots \mathfrak{x}_p$ is general if and only if $\mathfrak{z}$ is different from the $p - 1$ points $\mathfrak{y}_2, \ldots, \mathfrak{y}_p$ uniquely determined by $\mathfrak{x}_2, \ldots, \mathfrak{x}_p$. In that case (10) actually implies

$$\mathfrak{z}\mathfrak{x}_2 \cdots \mathfrak{x}_p = \mathfrak{a}\mathfrak{r}_2 \cdots \mathfrak{r}_p;$$

and, since $\mathfrak{a}$ is supposed different from $\mathfrak{x}_2, \ldots, \mathfrak{x}_p$, we have $\mathfrak{z} = \mathfrak{a}$ as well as, with proper indexing, $\mathfrak{x}_k = \mathfrak{r}_k$ $(k = 2, \ldots, p)$. Again, if $\mathfrak{z}$ coincides with one of the points $\mathfrak{y}_2, \ldots, \mathfrak{y}_p$, then the whole divisor $\mathfrak{z}\mathfrak{x}_2 \cdots \mathfrak{x}_p$ of degree p is special, so that (10) holds with prescribed $\mathfrak{a}$ and certain $\mathfrak{r}_2, \ldots, \mathfrak{r}_p$. This means that the equation $\varphi(\mathfrak{z}, \mathfrak{a}) = 0$ is satisfied only for $\mathfrak{z} = \mathfrak{a}, \mathfrak{y}_2, \ldots, \mathfrak{y}_p$. It remains to show that this statement accounts correctly for the multiplicities.

Since the function $\varphi(\mathfrak{p}, \mathfrak{a})$ does not vanish identically in $\mathfrak{p}$, it has on $\mathfrak{R}$, by Theorem 1 in the preceding section, exactly p zeros $\mathfrak{z}_1, \ldots, \mathfrak{z}_p$. Furthermore, by Theorem 2 in that section,

$$\sum_{k=1}^{p} w(\mathfrak{z}_k) \equiv w(\mathfrak{a}) + 2c - \sum_{k=2}^{p} w(\mathfrak{x}_k).$$

On the other hand, $\mathfrak{x}_2 \cdots \mathfrak{x}_p \mathfrak{y}_2 \cdots \mathfrak{y}_p$ is the divisor of the zeros of a differential of the first kind. This implies, by (9), the congruence

$$\sum_{k=2}^{p} w(\mathfrak{x}_k) + \sum_{k=2}^{p} w(\mathfrak{y}_k) \equiv q.$$

Therefore,

$$-w(\mathfrak{a}) + \sum_{k=1}^{p} w(\mathfrak{z}_k) - \sum_{k=2}^{p} w(\mathfrak{y}_k) \equiv 2c - q. \tag{11}$$

Here we supposed the given divisor $\mathfrak{x}_2 \cdots \mathfrak{x}_p$ general and the given point $\mathfrak{a}$ different from the points $\mathfrak{x}_2, \ldots, \mathfrak{x}_p$. We know that then the divisor $\mathfrak{y}_2 \cdots \mathfrak{y}_p$, complementary to $\mathfrak{x}_2 \cdots \mathfrak{x}_p$, is also general, and, apart from multiplicities, the p points $\mathfrak{z}_1, \ldots, \mathfrak{z}_p$ coincide with the p points $\mathfrak{a}, \mathfrak{y}_2, \ldots, \mathfrak{y}_p$. To prove coincidence of multiplicities, we choose the $p - 1$ points $\mathfrak{y}_2, \ldots, \mathfrak{y}_p$ distinct and the divisor $\mathfrak{y}_2 \cdots \mathfrak{y}_p$ general. The complementary divisor $\mathfrak{x}_2 \cdots \mathfrak{x}_p$ is then uniquely determined by the $\mathfrak{y}_2, \ldots, \mathfrak{y}_p$ and is likewise general. We choose $\mathfrak{a}$ distinct from the $2p - 2$ points $\mathfrak{x}_2, \ldots, \mathfrak{x}_p$ and $\mathfrak{y}_2, \ldots, \mathfrak{y}_p$. Since the p points $\mathfrak{a}, \mathfrak{y}_2, \ldots, \mathfrak{y}_p$ are distinct, we have in (11), after proper numbering, $\mathfrak{z}_1 = \mathfrak{a}$ and $\mathfrak{z}_k = \mathfrak{y}_k$ ($k = 2, \ldots, p$), while c and q are independent of $\mathfrak{a}, \mathfrak{y}_2, \ldots, \mathfrak{y}_p$. This proves the assertion

$$2c - q \equiv 0. \tag{12}$$

If we now assume only that the divisor $\mathfrak{x}_2 \cdots \mathfrak{x}_p$ is general, and $\mathfrak{a}$ is different from the points $\mathfrak{x}_2, \ldots, \mathfrak{x}_p$, then we can, at any rate, prescribe $\mathfrak{z}_1 = \mathfrak{a}$ and obtain from (11) and (12) the congruence

$$\sum_{k=2}^{p} w(\mathfrak{z}_k) \equiv \sum_{k=2}^{p} w(\mathfrak{y}_k).$$

But then, bearing in mind the fact that the divisor $\mathfrak{y}_2 \cdots \mathfrak{y}_p$ is also general, we have, by Abel's theorem,

$$\mathfrak{z}_2 \cdots \mathfrak{z}_p = \mathfrak{y}_2 \cdots \mathfrak{y}_p.$$

This completes the proof of Theorem 1.

After these preliminaries we can state the required generalization of (6).

Theorem 2: Let $f(\mathfrak{p})$ be a nonconstant function in the field K with the zeros $\mathfrak{a}_1, \ldots, \mathfrak{a}_m$ and the poles $\mathfrak{b}_1, \ldots, \mathfrak{b}_m$. We choose a general integral divisor $\mathfrak{x}_2 \cdots \mathfrak{x}_p$ of degree $p - 1$ such that $\mathfrak{x}_2, \ldots, \mathfrak{x}_p$ are different from the points $\mathfrak{a}_1, \ldots, \mathfrak{a}_m$ and $\mathfrak{b}_1, \ldots, \mathfrak{b}_m$. Then, after suitable choice of paths of integration, we have

$$\sum_{k=1}^{m} w(\mathfrak{a}_k) = \sum_{k=1}^{m} w(\mathfrak{b}_k), \tag{13}$$

and

$$f(\mathfrak{p}) = \delta \prod_{k=1}^{m} \frac{\Theta\left(w(\mathfrak{p}) - w(\mathfrak{a}_k) - c + \sum_{k=2}^{p} w(\mathfrak{x}_k)\right)}{\Theta\left(w(\mathfrak{p}) - w(\mathfrak{b}_k) - c + \sum_{k=2}^{p} w(\mathfrak{x}_k)\right)},$$

with the factor δ independent of $\mathfrak{p}$.

Proof: By Theorem 1, the zeros of the functions $\varphi(\mathfrak{p}, \mathfrak{a}_k)$ and $\varphi(\mathfrak{p}, \mathfrak{b}_k)$ $(k = 1, \ldots, m)$ on the Riemann surface $\mathfrak{R}$ are precisely $\mathfrak{a}_k, \mathfrak{y}_2, \ldots, \mathfrak{y}_p$ and $\mathfrak{b}_k, \mathfrak{y}_2, \ldots, \mathfrak{y}_p$, where $\mathfrak{y}_2 \cdots \mathfrak{y}_p$ is the divisor complementary to $\mathfrak{x}_2 \cdots \mathfrak{x}_p$. Hence the quotient of $\varphi(\mathfrak{p}, \mathfrak{a}_k)$ and $\varphi(\mathfrak{p}, \mathfrak{b}_k)$ has on $\mathfrak{R}$ just the simple zero $\mathfrak{a}_k$ and just the simple pole $\mathfrak{b}_k$. This implies that the product of these m quotients, viewed as a function of $\mathfrak{p}$ on $\mathfrak{R}$ has exactly the same zeros and poles as the given function $f(\mathfrak{p})$ in K.

If $\mathfrak{p}$ traverses a closed circuit on $\mathfrak{R}$, then, by Theorem 2, Section 9, $\varphi(\mathfrak{p}, \mathfrak{a})$ is multiplied by the factor

$$\gamma(\mathfrak{a}) = \varepsilon\left(-g'Zg - 2g'\left\{w(\mathfrak{p}) - w(\mathfrak{a}) - c + \sum_{k=2}^{p} w(\mathfrak{x}_k)\right\}\right),$$

where g denotes a suitable integer column independent of $\mathfrak{a}$. Then

$$\prod_{k=1}^{m} \frac{\gamma(\mathfrak{a}_k)}{\gamma(\mathfrak{b}_k)} = \varepsilon\left(2g'\left\{\sum_{k=1}^{m} w(\mathfrak{a}_k) - \sum_{k=1}^{m} w(\mathfrak{b}_k)\right\}\right). \tag{15}$$

Using Abel's theorem we see that, with suitably chosen paths of integration, (13) is satisfied. Now, by (15), the product on the right-hand side of (14) is a single valued meromorphic function on $\mathfrak{R}$, and, as such, belongs to the field K. Since this function has exactly the same zeros and poles as $f(\mathfrak{p})$, it differs from the latter function only by a factor δ independent of $\mathfrak{p}$. This completes the proof of the theorem.

Theorem 2 yields a simple representation of the functions in K in terms of theta quotients, which for $p = 1$ reduces to the uniformization by means of elliptic functions. We note, however, that the factor δ in (14) depends on the parameters $\mathfrak{x}_2, \ldots, \mathfrak{x}_p$. The nature of this dependence is made precise by the following theorem.

Theorem 3: Let $F(\mathfrak{p})$ be a nonconstant function in the field K with zeros $\mathfrak{a}_1, \ldots, \mathfrak{a}_m$ and poles $\mathfrak{b}_1, \ldots, \mathfrak{b}_m$ so that, with $w(\mathfrak{a}_k) = a_k$ and $w(\mathfrak{b}_k) = b_k$ $(k = 1, \ldots, m)$ and suitable choice of paths of integration, we have the relation

$$\sum_{k=1}^{m} a_k = \sum_{k=1}^{m} b_k.$$

If we put

$$\sum_{k=1}^{p} w(\mathfrak{x}_k) = s$$

with p independently varying points $\mathfrak{x}_1, \ldots, \mathfrak{x}_p$, then

$$f(\mathfrak{x}_1) \cdots f(\mathfrak{x}_p) = \gamma \prod_{k=1}^{m} \frac{\Theta(s - a_k - c)}{\Theta(s - b_k - c)}, \tag{16}$$

where the factor γ is independent of $\mathfrak{x}_1, \ldots, \mathfrak{x}_p$.

Proof: Let $\mathfrak{x} = \mathfrak{x}_1 \cdots \mathfrak{x}_p$ be a variable general divisor. Then for $l = 1, \ldots, p$ the divisor $\mathfrak{x}/\mathfrak{x}_l$ is also general. We put

$$q(s) = \prod_{k=1}^{m} \frac{\Theta(s - a_k - c)}{\Theta(s - b_k - c)}.$$

Now let us vary $\mathfrak{x}_l$ and take for the remaining $\mathfrak{x}_k$ ($k \neq l$) fixed points distinct from $\mathfrak{a}_1, \ldots, \mathfrak{a}_m$ and $\mathfrak{b}_1, \ldots, \mathfrak{b}_m$. Then, by Theorem 2, the quotient $f(\mathfrak{x}_l)/q(s)$ does not depend on $\mathfrak{x}_l$. If all the p points $\mathfrak{x}_1, \ldots, \mathfrak{x}_p$ vary independently, then the quotient

$$\frac{f(\mathfrak{x}_1) \cdots f(\mathfrak{x}_p)}{q(s)} = \gamma$$

does not depend on $\mathfrak{x}_1, \ldots, \mathfrak{x}_p$. The assertion follows.

Let $\mathfrak{x}_1, \ldots, \mathfrak{x}_p$ in (16) denote independent variables. If we replace $\mathfrak{x}_1, \ldots, \mathfrak{x}_p$ with definite points then (16) remains true, provided that the points in question are different from $\mathfrak{b}_1, \ldots, \mathfrak{b}_m$ and the divisor $\mathfrak{x}_1 \cdots \mathfrak{x}_p$ is general. On the other hand, the right-hand side of (16) loses meaning if we replace $\mathfrak{x}_1 \cdots \mathfrak{x}_p$ with a particular divisor; for, then $\Theta(s - a_k - c)$ and $\Theta(s - b_k - c)$ ($k = 1, \ldots, m$) all vanish. These remarks must be remembered when one tries to determine the factor γ by means of a definite choice of $\mathfrak{x}_1, \ldots, \mathfrak{x}_p$.

12. Jacobi–Abel functions

According to the investigations in Chapter 3, the automorphic functions are obtained by uniformization of the algebraic function fields of genus $p > 1$, whereas in the case $p = 1$ the uniformizing parameter is furnished by an elliptic integral of the first kind. For $p > 1$ no simple connection exists between the uniformizing parameter and the integrals of the first kind. As was made clear in Section 8, in the case $p > 1$ the inverse function of a single integral of the first kind is not single-valued in the large. We recall that the problem successfully investigated there, namely the *Jacobi inversion problem*, dealt with the inversion of p sums of p integrals of the first kind

$$s_k = \sum_{l=1}^{p} \int_{\mathfrak{a}_l}^{\mathfrak{x}_l} dw_k \qquad (k = 1, \dots, p).$$

We were given the column s with entries $s_1, \dots, s_p$ and were required to find the divisor $\mathfrak{x}_1 \cdots \mathfrak{x}_p$. The solution of the inversion problem is found in Theorems 1, 2, and 4 in Section 10; it is given in implicit form by means of the zeros of the theta function. Now we propose to give an explicit solution by expressing every symmetric rational function of $\mathfrak{x}_1, \dots, \mathfrak{x}_p$ rationally with constant coefficients in terms of functions $\Theta(s + a)$, where for a we choose suitable constant columns. For the products $f(\mathfrak{x}_1) \cdots f(\mathfrak{x}_p)$ this was already done in Theorem 3 of the preceding section.

Let Z again be the period matrix associated with a fixed canonical dissection of $\mathfrak{R}$ and a fixed normal basis $w_1, \dots, w_p$, and let $\Theta(s) = \Theta(Z, s)$ be the corresponding theta function. If $a_1, \dots, a_m$ are constant columns, then we call

$$\eta(s) = \Theta(s + a_1) \cdots \Theta(s + a_m)$$

a *theta product* and m its *order*. If we go from s to a congruent column

$$s^* = s + h + Zg \equiv s$$

with integral g and h, then, by Theorem 2 in Section 9, we have the relation

$$\eta(s^*) = \mu\eta(s), \qquad \mu = \varepsilon\left(-mg'Zg - 2mg's - 2g'\sum_{k=1}^{m} a_k\right). \tag{1}$$

We call $\mu = \mu_\omega$ the *multiplier* of $\eta(s)$ associated with the period $\omega = h + Zg$. We call two theta products

$$\eta_1(s) = \eta(s), \qquad \eta_2(s) = \Theta(s + b_1) \cdots \Theta(s + b_n)$$

equivariant if the two associated multipliers μ_ω coincide for every period. By (1), this happens if and only if the quotient $\eta_1(s)/\eta_2(s)$ viewed as a function of s has all the ω as periods. Again, this happens if and only if the two orders m and n are equal, and, in addition, for a suitable integer column d we have the relation

$$\sum_{r=1}^{m} a_r = d + \sum_{r=1}^{m} b_r. \tag{2}$$

We also admit the trivial case $m = 0$, and then define $\eta(s) = 1$.

Now let $f(\mathfrak{p})$ be a nontrivial field function with the zeros $\mathfrak{a}_1, \dots, \mathfrak{a}_m$ and the poles $\mathfrak{b}_1, \dots, \mathfrak{b}_m$; that is,

$$(f) = \frac{\mathfrak{a}_1 \cdots \mathfrak{a}_m}{\mathfrak{b}_1 \cdots \mathfrak{b}_m}.$$

For a suitable choice of the paths of integration, we then have, by Abel's theorem,

$$\sum_{r=1}^{m} w(\mathfrak{a}_r) = \sum_{r=1}^{m} w(\mathfrak{b}_r),$$

so that the columns

$$-w(\mathfrak{a}_r) - c = a_r, \qquad -w(\mathfrak{b}_r) - c = b_r \qquad (r = 1, \ldots, m)$$

satisfy the condition (2). It follows that the two theta products

$$\eta_1(s) = \prod_{r=1}^{m} \Theta(s - w(\mathfrak{a}_r) - c), \qquad \eta_2(s) = \prod_{r=1}^{m} \Theta(s - w(\mathfrak{b}_r) - c)$$

are equivariant and have the same order m. With variable $\mathfrak{x}_1, \ldots, \mathfrak{x}_p$ we put

$$s = \sum_{k=1}^{p} w(\mathfrak{x}_k). \tag{3}$$

By Theorem 3 in the preceding section, we have

$$f(\mathfrak{x}_1) \cdots f(\mathfrak{x}_p) = \gamma q(s), \qquad q(s) = \frac{\eta_1(s)}{\eta_2(s)} \tag{4}$$

with a suitably chosen constant γ.

Theorem 1: For every symmetric rational function $f(\mathfrak{x}_1, \ldots, \mathfrak{x}_p)$ of the p variable points $\mathfrak{x}_1, \ldots, \mathfrak{x}_p$ there are certain finitely many equivariant theta products $\eta_1(s), \ldots, \eta_h(s)$ and constants $c_1, \ldots, c_h, d_1, \ldots, d_h$ satisfying an identity

$$f(\mathfrak{x}_1, \ldots, \mathfrak{x}_p) = \frac{c_1\eta_1(s) + \cdots + c_h\eta_h(s)}{d_1\eta_1(s) + \cdots + d_h\eta_h(s)} \tag{5}$$

in which the denominator does not vanish identically.

Proof: Let $f(\mathfrak{p})$ be a nonconstant function in K and λ a constant. With variable $\mathfrak{x}_1, \ldots, \mathfrak{x}_p$ we have, in analogy to (4), a representation

$$\prod_{k=1}^{p} (\lambda + f(\mathfrak{x}_k)) = \gamma(\lambda) q(s, \lambda), \qquad q(s, \lambda) = \frac{\eta(s, \lambda)}{\eta(s)}, \tag{6}$$

where $\eta(s) = \eta_2(s)$, as defined above, and $\eta(s, \lambda)$ is a theta product equivariant with $\eta(s)$ whose factors are determined by the zeros of the function $\lambda + f(\mathfrak{p})$. Also, $\gamma(\lambda)$ is a number depending on λ. Multiplying the p factors on the left side of (6), we obtain the polynomial

$$\prod_{k=1}^{p} (\lambda + f(\mathfrak{x}_k)) = E_0\lambda^p + E_1\lambda^{p-1} + \cdots + E_p \tag{7}$$

with the coefficients

$$E_0 = 1, \qquad E_1 = f(\mathfrak{x}_1) + \cdots + f(\mathfrak{x}_p), \ldots, \qquad E_p = f(\mathfrak{x}_1) \cdots f(\mathfrak{x}_p).$$

If we substitute in it successively for λ the numbers $0, 1, \ldots, p$, then we obtain from (6) and (7) altogether $p + 1$ linear equations in the $p + 1$ unknowns $E_0, E_1, \ldots, E_p$ whose determinant is a Vandermonde determinant different from 0. Hence 1 and the p elementary symmetric polynomials in $f(\mathfrak{x}_1), \ldots, f(\mathfrak{x}_p)$ are quotients of homogeneous linear functions of $\eta(s, 0)$, $\eta(s, 1), \ldots, \eta(s, p)$ and the function $\eta(s)$. In particular, this representation holds for the sum

$$E_1 = f(\mathfrak{x}_1) + \cdots + f(\mathfrak{x}_p),$$

where $f(\mathfrak{p})$ may be constant.

Now let $f(\mathfrak{x}_1, \ldots, \mathfrak{x}_p)$ be any symmetric rational function of $\mathfrak{x}_1, \ldots, \mathfrak{x}_p$, that is, symmetric and rational in the p pairs of coordinates x_k, y_k $(k = 1, \ldots, p)$ of the points $\mathfrak{x}_k$. By forming the average over all the $p!$ permutations of these p pairs we can write $f(\mathfrak{x}_1, \ldots, \mathfrak{x}_p)$ as a quotient of symmetric polynomials. It is now clear that it is sufficient to prove Theorem 1 for only such polynomials; this is so because a product of theta products is again a theta product. We regard the x_k, y_k $(k = 1, \ldots, p)$ temporarily as $2p$ free variables and form the sum

$$\sigma_{qr} = \sum_{k=1}^{p} x_k^q y_k^r \qquad (q, r = 0, 1, 2, \ldots).$$

We are to show then that the polynomial $f(\mathfrak{x}_1, \ldots, \mathfrak{x}_p)$ can be expressed in an integral rational manner with constant coefficients in terms of finitely many σ_{qr}.

This is obvious for $p = 1$. We use induction on p, and assume that for a value of $p > 1$ every symmetric polynomial in the $p - 1$ pairs x_k, y_k $(k = 2, \ldots, p)$ can be expressed in an integral rational manner in terms of the corresponding sums

$$s_{qr} = \sum_{k=2}^{p} x_k^q y_k^r \qquad (q, r = 0, 1, 2, \ldots).$$

If we order $f(\mathfrak{x}_1, \ldots, \mathfrak{x}_p)$ in powers of x_1 and y_1 we obtain the decomposition

$$f(\mathfrak{x}_1, \ldots, \mathfrak{x}_p) = \sum_{k,l} \varphi_{kl} x_1^k y_1^l \tag{8}$$

into finitely many summands whose coefficients φ_{kl} are polynomials in the $p - 1$ pairs x_k, y_k $(k = 2, \ldots, p)$. In view of the symmetry of $f(\mathfrak{x}_1, \ldots, \mathfrak{x}_p)$ in all the p pairs, every coefficient φ_{kl} must be symmetric in the last $p - 1$ pairs, and so, by the induction assumption, must be a polynomial in the s_{qr}. In view of the relation

$$s_{qr} = \sigma_{qr} - x_1^q y_1^r,$$

(8) yields a new decomposition

$$f(\mathfrak{x}_1, \ldots, \mathfrak{x}_p) = \sum_{k,l} \psi_{kl} x_1^k y_1^l \tag{9}$$

whose coefficients ψ_{kl} are all polynomials in the σ_{qr}. Because of the symmetry of $f(\mathfrak{x}_1, \ldots, \mathfrak{x}_p)$ we can interchange in (9) the pairs x_1, y_1 and x_t, y_t ($t = 1, \ldots, p$). By forming the arithmetic mean we obtain from this

$$pf(\mathfrak{x}_1, \ldots, \mathfrak{x}_p) = \sum_{k,l} \psi_{kl}\sigma_{kl},$$

which completes the induction argument.

If we put, in particular,

$$f(\mathfrak{x}_k) = x_k^q y_k^r \qquad (k = 1, \ldots, p),$$

then we obtain

$$\sigma_{qr} = f(\mathfrak{x}_1) + \cdots + f(\mathfrak{x}_p),$$

and for this the assertion of Theorem 1 is already established. But then this assertion must be true for every polynomial in finitely many σ_{qr} and, therefore, quite generally, for the polynomial $f(\mathfrak{x}_1, \ldots, \mathfrak{x}_p)$. This completes the proof.

The quotient of two equivariant theta products, and, more generally, the function of s on the right side of (5) has all the columns ω belonging to the lattice Ω as periods. On the other hand, it is meromorphic. While postponing more detailed clarification to the next chapter, we define, in connection with our present objective, a *meromorphic function $q(s)$ of p complex variables* $s_1, \ldots, s_p$ as a quotient of two entire functions

$$q(s) = \frac{g(s)}{h(s)}, \tag{10}$$

with the denominator $h(s)$ not identically zero. If $j(s)$ is any nontrivial entire function, then, putting

$$g_1(s) = j(s)g(s), \qquad h_1(s) = j(s)h(s), \tag{11}$$

we have

$$q(s) = \frac{g_1(s)}{h_1(s)}. \tag{12}$$

We shall see in the next chapter that it is possible to determine $g(s)$ and $h(s)$ in (10) so that, conversely, a decomposition (12) with entire functions $g_1(s)$ and $h_1(s)$ implies (11), with a suitable entire function $j(s)$; such a pair $g(s)$, $h(s)$ is called *relatively prime*. Given a lattice Ω we define an *abelian function* as a meromorphic function for which all ω in Ω are periods. Later we will admit arbitrary lattices, that is, lattices generated by $2p$ columns linearly independent over the reals. In the present section, however, we adhere to the assumption that Ω consists of the periods of a normal basis of integrals of the first kind for a given function field K. For this reason it is more accurate to call the functions we are dealing with here *Jacobi-Abel functions.* Incidentally, the general abelian functions were introduced and studied long after Abel's death.

It is clear from the definition that the Jacobi-Abel functions form a field A. By Theorem 1, the substitution (3) carries every symmetric rational function in the variable points $\mathfrak{x}_1, \ldots, \mathfrak{x}_p$ into an element of A. We now show that the converse of this statement is true; so that, in particular, every Jacobi-Abel function can be represented in the form (5) as a quotient of two sums of equivariant theta products.

Theorem 2: The substitution

$$s = \sum_{k=1}^{p} w(\mathfrak{x}_k) \tag{13}$$

carries every Jacobi-Abel function into a symmetric rational function of $\mathfrak{x}_1, \ldots, \mathfrak{x}_p$.

Proof: The proof is in four steps.

I. Let $q(s)$ be a Jacobi-Abel function and (10) its representation as quotient of coprime entire functions. We choose a general point $s = s^*$ at which, in addition, $h(s^*) \neq 0$. (13) associates uniquely with s^* a general divisor $\mathfrak{x}^* = \mathfrak{x}_1^* \cdots \mathfrak{x}_p^*$ of degree p. On the Riemann surface $\mathfrak{R}$ we choose neighborhoods $\mathfrak{U}_1, \ldots, \mathfrak{U}_p$ of the points $\mathfrak{x}_1^*, \ldots, \mathfrak{x}_p^*$ so small that, viewed as a function of $\mathfrak{x}$, $h(s)$ is different from zero throughout the product neighborhood $\mathfrak{U}$ of $\mathfrak{x}^*$. We keep the $p - 1$ points $\mathfrak{x}_2, \ldots, \mathfrak{x}_p$ in the neighborhoods $\mathfrak{U}_2, \ldots, \mathfrak{U}_p$ fixed and leave the point $\mathfrak{x}_1$ with coordinates x, y variable. If we put

$$h(s) = h_0(\mathfrak{x}_1), \qquad g(s) = g_0(\mathfrak{x}_1), \qquad q(s) = q_0(\mathfrak{x}_1),$$

then the quotient

$$q_0(\mathfrak{x}_1) = \frac{g_0(\mathfrak{x}_1)}{h_0(\mathfrak{x}_1)} \tag{14}$$

is a regular function of $\mathfrak{x}_1$ in the neighborhood $\mathfrak{U}_1$ of $\mathfrak{x}_1^*$ on $\mathfrak{R}$. Since $h(s)$ and $g(s)$ are entire functions of s, and since, by (13), s viewed as a function of $\mathfrak{x}_1$ remains regular under arbitrary analytic continuation originating at $\mathfrak{x}_1^*$, the same is true for $h_0(\mathfrak{x}_1)$ and $g_0(\mathfrak{x}_1)$ as functions of $\mathfrak{x}_1$. Furthermore, as $\mathfrak{x}_1$ varies over a closed loop on $\mathfrak{R}$, s changes only by an additive period and $q(s)$, being an abelian function, remains invariant. It follows that $q_0(\mathfrak{x}_1)$ is single-valued and meromorphic on $\mathfrak{R}$, and thus a function in K. We note that in this argument $\mathfrak{x}_2, \ldots, \mathfrak{x}_p$ stayed fixed.

By (10), we have for every period ω the quotient representation

$$q(s) = \frac{g(s + \omega)}{h(s + \omega)},$$

where $g(s + \omega)$ and $h(s + \omega)$ are again entire functions of s. Since, on the other hand, $g(s)$ and $h(s)$ are relatively prime, it follows that

$$(15) \qquad g(s + \omega) = j_\omega(s)g(s), \qquad h(s + \omega) = j_\omega(s)h(s),$$

where $j_\omega(s)$ is an entire function which may also depend on ω. Then we also have

$$h(s) = j_\omega(s - \omega)h(s - \omega) = j_\omega(s - \omega)j_{-\omega}(s)h(s), \qquad j_\omega(s)j_{-\omega}(s + \omega) = 1,$$

so that $j_\omega(s)$ is nowhere zero. Hence, by (15), the two functions $h(s)$ and $h(s + \omega)$ have the same zeros. Concerning the function $h_0(\mathfrak{x}_1)$, we see that as the variables vary over a closed loop on $\mathfrak{R}$ every zero goes over into a zero of equal multiplicity, and the function itself without necessarily being single-valued on $\mathfrak{R}$ is multiplied by the factor $j_\omega(s)$. In view of the regularity of $h_0(\mathfrak{x}_1)$ and the compactness of $\mathfrak{R}$, this function has a certain number μ of zeros on $\mathfrak{R}$. If m is the order of $q_0(\mathfrak{x}_1)$, that is, if $q_0(\mathfrak{x}_1)$ has m poles on $\mathfrak{R}$, then, by (14), we have the estimate

$$(16) \qquad m \leqslant \mu.$$

We now effect a canonical dissection of $\mathfrak{R}$, so that no zeros of $h_0(\mathfrak{x}_1)$ lie on the cuts. Then μ is equal to the change in the function $(1/2\pi i) \log h_0(\mathfrak{x}_1)$ when $\mathfrak{x}_1$ traverses the boundary C of the canonically dissected surface $\mathfrak{R}^*$ once in the positive direction. Now $h(s)$ is also a regular function in each of the $p - 1$ parameters $\mathfrak{x}_2, \ldots, \mathfrak{x}_p$, and, in particular, is continuous in the latter. We can therefore suppose the neighborhoods $\mathfrak{U}_2, \ldots, \mathfrak{U}_p$ chosen so small that, for every choice of $\mathfrak{x}_2, \ldots, \mathfrak{x}_p$ in these neighborhoods, the function $h_0(\mathfrak{x}_1)$ is different from 0 on all of C. Since μ must then change continuously with $\mathfrak{x}_2, \ldots, \mathfrak{x}_p$ in these neighborhoods, and is, on the other hand, an integer, it follows that μ is constant there.

II. As in Section 1 of Chapter 2 (Volume I), we now introduce the individual sheets $\mathfrak{S}_1, \ldots, \mathfrak{S}_n$ of $\mathfrak{R}$ and denote by $q_1(x), \ldots, q_n(x)$ the corresponding branches of $q_0(\mathfrak{x}_1)$. If $P(y, x)$ has the same meaning as the polynomial $P(u, z)$ introduced there, and $y_1(x), \ldots, y_n(x)$ are the n branches of y, then, in analogy to the formula (8) of that section, we form the equality

$$(17) \qquad g(y, x) = \sum_{k=1}^{n} \frac{q_k(x)P(y, x)}{(y - y_k(x))P_y(y_k(x), x)}$$
$$= T_1(x)y^{n-1} + T_2(x)y^{n-2} + \cdots + T_n(x),$$

where, to begin with, y is an indeterminate independent of x. If we put in it $y = l$ with $l = 1, \ldots, n$ then we obtain $T_1(x), \ldots, T_n(x)$ as n homogeneous linear functions of $g(l, x)$ $(l = 1, \ldots, n)$ with constant coefficients. Since $T_1(x), \ldots, T_n(x)$ are single-valued and meromorphic on the schlicht x-sphere $\mathfrak{K}$, they must be rational functions of x. Now we can choose n constants $\lambda_1, \ldots, \lambda_n$ such that the rational function

$$T(x) = \lambda_1 T_1(x) + \cdots + \lambda_n T_n(x)$$

has as reduced denominator just the common denominator of $T_1(x), \ldots, T_n(x)$. By (17), the number of poles of $T(x)$ on $\mathfrak{K}$ including multiplicities is at most equal to $m + c$, where c is determined by the number of poles of the functions appearing in (17) as factors of $q_1(x), \ldots, q_n(x)$ and is therefore independent of $q(s)$.

If y denotes again the algebraic function of x defined by the equation $P(y, x) = 0$, then

$$q_0(\mathfrak{x}_1) = T_1(x)y^{n-1} + \cdots + T_n(x) \tag{18}$$

is an explicit rational function of x and y such that the degrees in x of its numerator and denominator are bounded from above by $\mu + c$. The main point in this statement is that, while the coefficients of the rational functions $T_1(x), \ldots, T_n(x)$ depend to some extent on the points $\mathfrak{x}_2, \ldots, \mathfrak{x}_p$, the relevant degrees in x remain bounded as long as $\mathfrak{x}_2, \ldots, \mathfrak{x}_p$ lie in the neighborhoods $\mathfrak{U}_2, \ldots, \mathfrak{U}_p$ of $\mathfrak{x}_2^*, \ldots, \mathfrak{x}_p^*$. It is clear that in the previous argument we could have singled out as variable in place of $\mathfrak{x}_1$ one of the remaining points $\mathfrak{x}_2, \ldots, \mathfrak{x}_p$, and arrived at the corresponding statement concerning the boundedness of degree.

III. We now prove the theorem by induction. Let r be an integer in the sequence 1 to p and suppose that, if r points $\mathfrak{x}_1, \ldots, \mathfrak{x}_r$ are variable and the remaining $p - r$ points $\mathfrak{x}_{r+1}, \ldots, \mathfrak{x}_p$ are restricted to sufficiently small neighborhoods $\mathfrak{U}_{r+1}, \ldots, \mathfrak{U}_p$ of $\mathfrak{x}_{r+1}^*, \ldots, \mathfrak{x}_p^*$, then there exists a representation

$$q(s) = q_0(\mathfrak{x}_1, \ldots, \mathfrak{x}_r) = \frac{\sigma_r}{\tau_r}, \tag{19}$$

where σ_r is a polynomial in the coordinates x_k, y_k $(k = 1, \ldots, r)$ of $\mathfrak{x}_1, \ldots, \mathfrak{x}_r$ and τ_r is a polynomial in $x_1, \ldots, x_r$ alone. The coefficients in these polynomials depend also on $\mathfrak{x}_{r+1}, \ldots, \mathfrak{x}_p$; moreover, we suppose the degrees of σ_r and τ_r relative to all the variables $x_1, \ldots, x_r$ bounded, and, in addition, each of the variables $y_1, \ldots, y_r$ appears in σ_r to degree at most $n - 1$. This statement was proved above for the case $r = 1$. Now let $1 \leqslant r \leqslant p - 1$. If h is an upper bound independent of $\mathfrak{x}_{r+1}, \ldots, \mathfrak{x}_p$ for the degrees of σ_r and τ_r as functions of each of the r variables $x_1, \ldots, x_r$, then we form all the power products $x_1^{k_1} \cdots x_r^{k_r} y_1^{l_1} \cdots y_r^{l_r}$ with $k_1, \ldots, k_r = 0, 1, \ldots, h$ and $l_1, \ldots, l_r = 0, 1, \ldots, n - 1$ and denote them in definite order by $\varphi_1, \ldots, \varphi_a$, with $a = (h + 1)^r n^r$. Furthermore, we denote all the corresponding power products $x_1^{k_1} \cdots x_r^{k_r}$ by $\psi_1, \ldots, \psi_b$ with $b = (h + 1)^r$. Then

$$\sigma_r = \lambda_1 \varphi_1 + \cdots + \lambda_a \varphi_a, \qquad \tau_r = \mu_1 \psi_1 + \cdots + \mu_b \psi_b \tag{20}$$

with certain coefficients $\lambda_1, \ldots, \lambda_a$ and $\mu_1, \ldots, \mu_b$, the latter not all zero, which depend on $\mathfrak{x}_{r+1}, \ldots, \mathfrak{x}_p$ alone.

Now we apply (18) with $\mathfrak{x}_{r+1}$ in place of $\mathfrak{x}_1$ and corresponding to the case $r = 1$ of (19) we obtain a representation

$$q(s) = \frac{\sigma}{\tau}, \tag{21}$$

where $\sigma = \sigma(x_{r+1}, y_{r+1})$ and $\tau = \tau(x_{r+1})$ are polynomials whose coefficients depend on the remaining $p - 1$ points $\mathfrak{x}_k$ ($k \neq r + 1$). We now make the assumption that $\mathfrak{x}_1, \ldots, \mathfrak{x}_r$ also lie in sufficiently small neighborhoods $\mathfrak{U}_1, \ldots, \mathfrak{U}_r$ of $\mathfrak{x}_1^*, \ldots, \mathfrak{x}_r^*$. (19) and (21) imply the equality

$$\tau \sigma_r - \sigma \tau_r = 0, \tag{22}$$

which, by (20), is linear homogeneous in the $a + b$ quantities $\lambda_1, \ldots, \lambda_a$ and $\mu_1, \ldots, \mu_b$. As coefficients of these linear equations we have the expressions $\tau \varphi_k$ ($k = 1, \ldots, a$) and $-\sigma \psi_k$ ($k = 1, \ldots, b$), which we denote, in some fixed order, by χ_k ($k = 1, \ldots, a + b$). To emphasize the dependence on $\mathfrak{x}_1, \ldots, \mathfrak{x}_r$ we shall write, more accurately $\chi_k(\mathfrak{x}_0)$, where $\mathfrak{x}_0$ is the divisor $\mathfrak{x}_1 \cdots \mathfrak{x}_r$. With $d = a + b$, we form d such independently variable divisors $\mathfrak{x}_0 = \mathfrak{x}_0^{(1)}, \ldots, \mathfrak{x}_0^{(d)}$, and, out of the corresponding elements

$$\chi_{kl} = \chi_k(\mathfrak{x}_0^{(l)}) \qquad (k, l = 1, \ldots, d),$$

we form the matrix

$$\mathfrak{M} = (\chi_{kl}).$$

Here we note that χ_{kl} depends on the points $\mathfrak{x}_{r+1}, \ldots, \mathfrak{x}_p$, the dependence being integral rational with bounded degrees of the coordinates x_{r+1}, y_{r+1} of the point $\mathfrak{x}_{r+1}$ in $\mathfrak{U}_{r+1}$. Let ρ be the rank of $\mathfrak{M}$ for a fixed choice of $\mathfrak{x}_{r+2}, \ldots, \mathfrak{x}_p$ in $\mathfrak{U}_{r+2}, \ldots, \mathfrak{U}_p$ and variable $\mathfrak{x}_0^{(1)}, \ldots, \mathfrak{x}_0^{(d)}, \mathfrak{x}_{r+1}$. Since (22) holds with $\mu_1, \ldots, \mu_b$ not all zero, we must have, in any case, $0 \leqslant \rho \leqslant d - 1$.

IV. Since $\mathfrak{x}_0^{(1)}, \ldots, \mathfrak{x}_0^{(d)}$ are independent variables, the first ρ columns of $\mathfrak{M}$ already contain a ρ-rowed subdeterminant which does not vanish identically and which we may suppose constructed out of the first ρ rows. If we write again $\mathfrak{x}_0$ for $\mathfrak{x}_0^{(\rho+1)}$ and expand the subdeterminant formed with the first $\rho + 1$ rows of $\mathfrak{M}$ by the last column, then we obtain an equality of the form

$$\nu_1 \chi_1(\mathfrak{x}_0) + \cdots + \nu_{\rho+1} \chi_{\rho+1}(\mathfrak{x}_0) = 0 \tag{23}$$

identically in $\mathfrak{x}_0, \mathfrak{x}_0^{(1)}, \ldots, \mathfrak{x}_0^{(\rho)}$ and $\mathfrak{x}_{r+1}$. Here the coefficients $\nu_1, \ldots, \nu_{\rho+1}$ depend on $\mathfrak{x}_0^{(1)}, \ldots, \mathfrak{x}_0^{(\rho)}$, and, in an integral rational manner with bounded degrees, on x_{r+1}, y_{r+1}, but are independent of $\mathfrak{x}_0$. If we put in (23) the original expressions for the χ_k, then we obtain the equality

$$(24) \qquad \tau(\alpha_1\varphi_1 + \cdots + \alpha_a\varphi_a) - \sigma(\beta_1\psi_1 + \cdots + \beta_b\psi_b) = 0,$$

where the coefficients $\alpha_1, \ldots, \alpha_a$ and $\beta_1, \ldots, \beta_b$ are formed from the values $0, \nu_1, \ldots, \nu_{\rho+1}$. They are not all zero, since, in particular, $\nu_{\rho+1}$ is different from zero. If $\beta_1, \ldots, \beta_b$ were all zero, then, contrary to the irreducibility of the equation $P(y, x) = 0$, we would obtain in (24) a linear homogeneous dependence relation connecting the power products $\varphi_1, \ldots, \varphi_a$. Therefore, (21) and (24) imply

$$(25) \qquad q(s) = \frac{\alpha_1\varphi_1 + \cdots + \alpha_a\varphi_a}{\beta_1\psi_1 + \cdots + \beta_b\psi_b}.$$

Here we can now replace $\mathfrak{x}_0^{(1)}, \ldots, \mathfrak{x}_0^{(\rho)}$ with fixed divisors so that the denominator on the right does not vanish identically in the $r + 1$ points $\mathfrak{x}_1, \ldots, \mathfrak{x}_r, \mathfrak{x}_{r+1}$. After application of the equation $P(y_{r+1}, x_{r+1}) = 0$ and extension by means of a suitable polynomial in x_{r+1}, the numerator and denominator in (25) yield the required polynomials σ_{r+1} and τ_{r+1}. The restriction of $\mathfrak{x}_1, \ldots, \mathfrak{x}_{r+1}$ to the neighborhoods $\mathfrak{U}_1, \ldots, \mathfrak{U}_{r+1}$ may be subsequently dropped, since, for fixed $\mathfrak{x}_{r+2}, \ldots, \mathfrak{x}_p$, both sides of (25) are meromorphic for arbitrarily varying $\mathfrak{x}_1, \ldots, \mathfrak{x}_{r+1}$. This completes the induction argument.

For $r = p$ we obtain $q(s)$ from (19) as a rational function of x_k, y_k $(k = 1, \ldots, p)$. In view of the symmetry of s in the p variables $\mathfrak{x}_1, \ldots, \mathfrak{x}_p$, we can apply to the formula so obtained, all the permutations of the p pairs x_k, y_k and get the assertion of the theorem by forming the arithmetic mean.

Theorem 2 yields the important insight that the Jacobi-Abel functions have the same significance for the symmetric rational functions of p points of a Riemann surface of genus $p > 1$ as do the elliptic functions for the case $p = 1$. The following three theorems pertaining to Jacobi-Abel functions also yield, for $p = 1$, important properties of elliptic functions.

Theorem 3: The field of Jacobi–Abel functions is an algebraic function field of degree of transcendence p.

Proof: By Theorem 1, the p elementary symmetric functions in $x_1, \ldots, x_p$ are, as functions of s, Jacobi-Abel functions and it is clear that they are not related by an algebraic equation with constant coefficients. Now let $f_1, \ldots, f_p$ be p algebraically independent elements of the field A, and f an additional Jacobi-Abel function. By Theorem 2, there are $2p + 2$ symmetric polynomials $S, S_1, \ldots, S_p$ and $T, T_1, \ldots, T_p$ in the p pairs x_k, y_k $(k = 1, \ldots, p)$

such that, after application of the substitution (13), we have the $p + 1$ equations

$$S - Tf = 0, \qquad S_k - T_k f_k = 0 \qquad (k = 1, \ldots, p). \tag{26}$$

To these are added the p further algebraic equations

$$P(y_k, x_k) = 0 \qquad (k = 1, \ldots, p). \tag{27}$$

If we view f and $f_1, \ldots, f_p$ as unknowns, then they are connected by an algebraic equation with constant coefficients which we obtain through elimination of the $2p$ variables x_k, y_k $(k = 1, \ldots, p)$ by means of repeated formation of resultants of the $2p + 1$ equations (26) and (27). Since, for given functions $f_1, \ldots, f_p$, all but the first of these equations are also given and the first is linear in f, the degree of f in the resultants remains bounded. It follows that every element f in A satisfies an algebraic equation of bounded degree whose coefficients are polynomials in $f_1, \ldots, f_p$.

Well-known algebraic arguments justify the conclusion that the field A of all Jacobi-Abel functions is a simple algebraic extension of the field A_0 of the algebraic functions of $f_1, \ldots, f_p$. We can therefore choose a function $f = f_0$ such that A arises from A_0 by adjunction of f_0, and f_0 is related to $f_1, \ldots, f_p$ by an irreducible algebraic equation with constant complex coefficients. This completes the proof.

Theorem 4: The p partial derivatives $f_{s_1}, \ldots, f_{s_p}$ of every Jacobi-Abel function f are again Jacobi-Abel functions and are connected with f by means of an algebraic equation with constant coefficients.

Proof: We see from the definition (10) that every partial derivative of a meromorphic function of p variables is again a meromorphic function. By differentiating the equality $f(s + \omega) = f(s)$ we see that ω is a period for each of the derivatives $f_{s_1}, \ldots, f_{s_p}$, so that these derivatives and f all belong to A. By Theorem 3, the $p + 1$ functions $f, f_{s_1}, \ldots, f_{s_p}$ must, as elements of A, be algebraically dependent. This completes the proof.

For $p = 1$ Theorem 3 states that for a given period lattice any two elliptic functions are algebraically dependent; while Theorem 4 yields for every elliptic function an algebraic differential equation of the first order with constant coefficients. This follows already from Section 14 in Chapter 1 (Volume I). We now come to the corresponding generalization of the addition theorem for elliptic functions.

Theorem 5: Let s and t be two columns of p independent variables each, and let $f_0, f_1, \ldots, f_p$ be generators for the field of Jacobi-Abel functions. There exist $p + 1$ rational functions

$$R_k = R_k(u_0, \ldots, u_p; v_0, \ldots, v_p) \qquad (k = 0, \ldots, p)$$

of $2p + 2$ variables $u_0, \ldots, u_p$ and $v_0, \ldots, v_p$ with constant coefficients such that none of the denominators of $R_0, \ldots, R_p$ vanishes identically in s and t for $u_l = f_l(s)$, $v_l = f_l(t)$ $(l = 0, \ldots, p)$, and the equation

$$(28) \quad f_k(s + t) = R_k(f_0(s), \ldots, f_p(s); f_0(t), \ldots, f_p(t)) \qquad (k = 0, \ldots, p)$$

holds.

Proof: We can make use of the reasoning in the proof of Theorem 2 by putting $f_k(s + t) = q(s)$ for a fixed k in the sequence $0, \ldots, p$ and a fixed t. We then show by the previous argument that $f_k(s + t)$ can be expressed as a rational function of $f_0(s), \ldots, f_p(s)$, where the coefficients admittedly depend on t, but the degrees are bounded. If we also interchange s and t, then we can carry over in a natural way the reasoning relating to (19) and (21), and so complete the proof. The sketchy nature of the present argument is justified by the fact that in the next chapter (Volume III) we give a detailed proof along similar lines of the addition theorem for arbitrary abelian functions.

Another proof can be related to the addition theorem for integrals of the first kind stated in Theorem 2 of Section 7. To this end we use the formula (22) of that section and put

$$s = \int_{\mathfrak{y}}^{\mathfrak{x}} dw, \qquad t = \int_{\mathfrak{y}}^{\mathfrak{z}} dw, \qquad \mathfrak{y} = \mathfrak{p}_0 \cdots \mathfrak{p}_0 = \mathfrak{p}_0^p.$$

By Theorem 2 in the preceding section we obtain immediately the required assertion, provided that the indeterminate form $0/0$ does not arise on the right-hand side of the addition theorem for integrals of the first kind corresponding to (28) in consequence of the specialization $\mathfrak{y}_l = \mathfrak{p}_0$ $(l = 1, \ldots, p)$. The latter difficulty can be easily overcome if, with variable $\mathfrak{y}_l$, we pass successively and individually to the limit $\mathfrak{y}_l \to \mathfrak{p}_0$ for $l = 1, \ldots, p$. Conversely, using the sketched first proof of Theorem 5, it is possible to obtain again the addition theorem for integrals of the first kind.

In this section we studied the Jacobi-Abel functions, that is, the special abelian functions whose period lattice is formed out of the periods of the integrals of the first kind associated with a given compact Riemann surface. If we free ourselves from this restriction, then we arrive at the functions which will be treated extensively in the next chapter (Volume III).

CUMULATIVE INDEX

Roman numeral preceding page numbers indicates volume